Playing By The Rules

A Clear Distinction between Science and Religion

By Rev. Dr. Paul Puffe

Copyright © 2020 by Paul Puffe

Sentia Publishing Company has the exclusive rights to reproduce this work, to prepare derivative works from this work, to publicly distribute this work, to publicly perform this work, and to publicly display this work.

All rights reserved. No part of this publication may be reproduced, stored in a retrieval system, or transmitted, in any form or by any means, electronic, mechanical, photocopying, recording, or otherwise, without the prior written permission of the copyright owner.

Most Bible quotations in this text, unless otherwise specified, are taken from the *HOLY BIBLE, NEW INTERNATIONAL VERSION® NIV.* Copyright © 1973, 1978, 1984, 2011 by Biblica, Inc. Used by permission.

Some Bible quotations are taken from *GOD'S WORD Translation.* Scripture references are used by permission. Copyright 1995 by God's Word to the Nations Bible Society.

Printed in the United States of America
ISBN 978-1-7352971-9-4

Table of Contents

Prologue: The Author's Purpose v

Chapter One: Different Games, Different Rules 1

Chapter Two: Why Use the Label "Game"? 6

    The Distinction Between Natural and Supernatural 17

Chapter Three: The Game of Science and the Rules of Modern Science 23

    A Brief Review of the Creation of Modern Natural Science 23

    The Rules of the Game of Science 39

Chapter Four: The Rules of the Game of Religion 51

Chapter Five: The Importance of Sacred Scripture in Modern Religion 55

    A Brief Discussion of Types of Religion 55

        Ancient and Tribal Religions 56

        Religions that Focus on Mental Powers 57

        Social Religions 58

        Religions that Recognize both the Natural and the Supernatural 59

    A Brief Review of the Interpretation of Supernatural Revelation 61

        Interpreting Supernatural Actions 62

        Interpreting Supernatural Signs 64

    The Role of Scripture as Divine Revelation 69

Chapter Six: The Rules for Interpreting Written Revelation 77
    The Rules for Religion with Written Revelation 77
    The Role of Human Reason in Supernatural Interpretation ... 108

Chapter Seven: The Problem with Intelligent Design 119

Chapter Eight: Design and Evolution 128

Chapter Nine: Prayer, Probability, and the Problem with Irresponsible Prayer 153

Chapter Ten: Comments on the Genesis 1 Creation Story 164

Conclusion 193

Prologue: The Author's Purpose

The purpose of this book is to help high school and college students gain a clearer understanding of the distinction between the study of natural science and the study of religion. The world of the twentieth and early twenty-first centuries has witnessed great debate over the relationship between these two topics as the techniques of modern natural science have brought to light new knowledge that has changed our understanding of ancient history. By the label "ancient history" I refer to what is sometimes called "pre-history," that is, the story of the events on Earth (and in the larger universe) prior to the written records of humans. These events include the formation of the planet Earth and the development of life upon Earth, especially the development and growth of human life on Earth. Although in one technical sense you cannot have "history" until you have written records, it is still common to use the term "history" in a looser sense to refer to the events prior to the invention of writing, and to also use the term to refer to the events on Earth prior to the time that most scholars envision humans operating on Earth, such as the age of the dinosaurs.

Modern natural science has created this narrative of ancient history from the study of geology, paleontology, and astronomy, with the assistance of many other branches of natural science. But this relatively new narrative of ancient history describes a different picture than that which was presumed to be true for many centuries by Jews and Christians. The earlier picture had been conceived by studying the creation account found in Genesis 1-2, along with other Bible passages. The difference between these two pictures has resulted in academic conflict. Some people have used this new understanding of ancient history to argue that it establishes that the Bible is not an accurate source of information about ancient history, and therefore not an accurate source of information about any history, and thus is to be rejected as a valid source of any information, even about God. Others have argued that the picture developed from the study of the Bible must take priority, and that this picture therefore exposes the newer narrative from natural science as erroneous, needing to be corrected by harmonization with the picture developed from Biblical study. And third, some people have used the new understanding of ancient history developed from natural science to adjust their interpretation of the religious information provided in the Bible, and concluded that no tension between this new picture and the religious information in the Bible was necessary. This book argues for the last of these options: the results of modern natural science regarding ancient history create a picture that does not need to be placed in conflict with the Bible or in conflict with the religion developed from the Bible, and both sources of information are to be appreciated.

The method that this book will use to help gain a clearer understanding of the distinction between the study of natural science and the study of religion is a large analogy: the different fields of human scholarship are like different games, and each game has to be played with its own rules. Modern natural science will be defined as one game; the rules of modern science will be discussed, and the evolution of those rules in recent centuries will be explained. Religion will be defined as another game. Different types of religion will be discussed briefly, and then this book will focus on the tradition of religious reasoning that uses sacred scriptures as its source of information from the supernatural world. The rules for studying sacred scripture will be described. Following this foundation, the later chapters of this book will look at some of the topics where significant debate between science and religion has occurred. These topics will include the question of design and creation, the concept of evolution and its place in the modern science of biology, some of the questions associated with how humans should pray for supernatural help, and some of the debate over the interpretation of chapter one of the book of Genesis in the Bible.

I, the author of this book, am a Lutheran Christian, and a professor of theology at a Christian university. As a young man I wanted to be an astronaut, and so I attended the Massachusetts Institute of Technology, earning a Bachelor of Science degree in Aeronautical and Astronautical Engineering in 1975. But while in college I developed a greater appreciation for the role of religious information in providing guidance for the purpose of each human life. I began to participate in some of the various Christian fellowship groups at MIT, including serving in some leadership roles. My sense of God's direction for my life led me to enter Concordia Seminary in St. Louis after college graduation. I received an excellent education in theology, but again as I approached graduation from the seminary I found myself turning in a new direction. I had entered the seminary intending to study the New Testament and become a parish pastor. Instead I found myself studying the Old Testament, and looking at further graduate school when I finished my Master of Divinity degree in 1979. Thus I continued studying at the University of Michigan in the field of Near Eastern Studies, learning about the languages, literatures, and archaeological history of the ancient Middle Eastern world that made up the neighborhood in which the people of the Old Testament lived. During this period I also served as pastor of a small rural parish outside of Ann Arbor. I received a Master of Arts degree in 1983.

Then I found myself taking another unexpected turn: I shifted from the parish to a teaching position at Concordia University Texas, in Austin. Here I taught Biblical Hebrew, Old Testament studies, world religions, and other theology courses for over thirty years. While teaching at Concordia I also enrolled at the Austin Presbyterian Theological Seminary and earned a Doctor of Ministry degree in 2012 in Christian Education. This latest turn in my educational history helped me to tie all of my previous education together. In addition to studying the methods of conducting religious education, I also studied the history and philosophical background of theological education. I was able to look at the trends of theological education, why certain ideas were popular at different times, and how that had affected my own history and education. It is the combination of these experiences in studying science, theology, Near Eastern literature, and the history and methodology of theological education that makes me bold enough to offer this discussion on the relationship of the study of science and religion.

The content of this book is basically what scholars call *hermeneutics*. Hermeneutics is the study of interpretation. This field of study has a wide range of specialized areas: it functions in philosophy with the question of how we perceive and interpret reality; it functions in science with the question of how we interpret the results of scientific experiments; and it functions in theology with the question of how we interpret the sacred scriptures. The premise behind this book is that a clearer understanding about how we think about science and religion, and how we go about the process of interpreting the data from both fields, will result in less confusion and debate. We will be able to see how both fields can and should exist side by side, and how both contribute to a greater understanding of the world. Let me begin with a discussion of how different games have different rules.

Chapter One: Different Games, Different Rules

If you knew nothing about the games of chess and checkers, and someone was introducing you to them for the first time, the games might seem to be nearly identical to you. Both are played on the same board, a square containing 64 smaller squares of alternating colors, arranged in a pattern of 8 rows and 8 columns. A common color scheme for the checkers board is alternating red and black squares. (Chess boards may alternate white and black squares, or other color combinations are often used with decorative sets, e.g., gold and silver, but many people use the same red and black board for both games.) Both games involve placing playing pieces on the board, lining them up on opposite sides. The players move the pieces, one at a time in alternating turns, generally toward the opposite side of the board. In both games if certain pieces make it all the way to the other side, they change into other pieces. Both games have certain pieces that are called "kings." In both games a player seeks to capture the pieces of his opponent and remove them from the board. In both games there are moments when a piece can hop over the piece of the opponent. Both games call for the careful strategy of positioning the pieces so as to trap and capture the opponent's pieces. All these similarities can lead one to the conclusion that these games are nearly the same.

But it only takes a little experience with both games to realize how different they actually are. In the game of checkers, all the pieces on each side are the same at the beginning, and they are placed only on black squares; none of the pieces ever move onto a red square. In the game of chess, there are several different types of pieces which are labeled pawns, rooks, knights, bishops, queens, and kings, and they are placed on both the red and black squares. In the game of checkers all of the pieces move the same way, sliding forward one square diagonally until they might reach the other side of the board. If they do reach the other side, they become "kings," and they are allowed to move backward as well as forward. In checkers you capture the piece of an opponent by hopping over it, diagonally, to an open square on the other side. In chess you capture an opponent's piece by landing on it. In the game of chess the different types of pieces move each in their unique pattern. Pawns can only move forward, one square at a time (except for a special rule about the first move of each pawn). If a pawn reaches the other side of the board, it can change into any other piece except a king (in contrast to checkers). Knights are the only pieces in chess that can hop over another piece, and they move in any direction in a motion shaped like the letter L. Rooks can move either forward, backward, or sideways, but not diagonally, and they can move any number of squares you choose. Bishops move only diagonally, but any number of squares forward or backward. Queens can move any number of squares either forward, backward, sideways, or diagonally, and are considered the most powerful pieces in the game. Kings can only move one square at a turn, but in any direction. However, the king is the most important piece in a chess game: the game ends when one player makes the capture of his opponent's king inevitable.

    So while both checkers and chess are games of strategy and are played on the same board, they each have their own rules, and thus they require fairly different kinds of strategy. It is not uncommon for a person who is very good at playing one of these games to not be as good at playing the other.

    The purpose of this book is to compare the relationship between the academic fields of science and religion to the relationship between games such as chess and checkers, and to clarify that you must operate with different rules when working within these different academic fields, just as you use different rules when playing these different games. Both the fields of science and religion seek to learn what is true about the world and about life. But just as chess and checkers, for all their similarity, are played with different rules, so also the fields of science and religion use different rules, despite their common goal of seeking to learn truth. In this book I am going to refer to the academic fields of science and religion as games, and talk about the "game" of science and the "game" of religion. In Chapter Two I will say more about the value of using this label "game" when considering how these academic fields conduct their search for truth.

Before moving on to discussing the games of science and religion, let me note that what has been said about checkers and chess is also true of many other games. Various games can have many similarities, but each game has its own rules that must be followed. If you saw a narrow rectangular net tied between two trees in someone's yard, you might ask what that was for, and be told it was for playing volleyball. In volleyball two different teams, one on each side of the net, bounce a ball back and forth. Each team wants to prevent the ball from hitting the ground when it is on their side of the net, but to force it to hit the ground when it is in the territory of the opponent. If you then walked further down the street and found another net in another yard, you might ask to watch these new people play volleyball, only to be told that this net was for a game called badminton, and while it has several similarities with volleyball, it is not played with a ball at all, but instead it is played with a feathered cone called a shuttlecock that is struck with racquets.

As another example we can compare the game called football in the United States with the game called football in Europe or South America—the game called soccer in North

America. Both games use a similar rectangular field of grass, of about the same size, and both have teams of players that attempt to move the ball from one end of the field to the other, to score points by placing the ball in the proper goal location. But while both games have rules calling for kicking the ball in certain situations and for throwing it in certain other situations, those situations and rules are quite different from one game to the other. In soccer the ball can only be thrown by a player called a goalie, unless the ball is being returned to play after being kicked out of bounds, in which case any player can do the throw. Otherwise, every player can only kick the soccer ball, or bounce it with a part of his body that cannot include the arms or hands. In American football, on the other hand, kicking the ball with one's feet occurs much less frequently, and the ball is usually advanced by carrying it, or by throwing it by hand from one player to another. In American football the opposing team members are expected to have contact with each other. There is much bumping and pushing, and much of the game involves trying to tackle the person carrying the ball. But the rules of soccer do not permit carrying the ball, and do not permit tackling, or intentional bumping or shoving, but only a certain amount of accidental and incidental bumping and banging that takes place as the players all pursue kicking the ball around the field. Furthermore, the balls do not look identical: a soccer ball is a sphere, while in American football the ball is shaped as an oval with two pointed ends.

Two other sports that are very similar are baseball and softball. These games, which always use different size balls, are so similar that it is common in informal play in neighborhood games for the rest of the equipment to be interchanged: bats and mitts designed for softball are used with a baseball, and vice versa. But in league sports these games use different size fields, different size bats, and different size mitts. They also have different rules. For example, in league softball, a player on a base must remain on the base until the ball is in motion, whereas in baseball a player on a base is permitted to "lead off" a base and begin moving toward the next base prior to the pitcher releasing the ball.

As a final set of examples, let us consider the world of playing cards. The games of bridge and poker are played with the same deck of cards, but the rules are very different. The games of bridge and pinochle are more similar to each other in rules and strategy, but they are played with different decks of cards.

So for any game that you choose to play, it is important to know what the particular rules are. In each game you must play by the appropriate rules. If you try to use the rules of chess when moving a checker piece, or you try to use the rules of checkers when capturing a chess piece, you destroy the structure of the game, and you destroy the entire interaction involved in playing a game with someone. If a person familiar with baseball tries to lead off and steal second base during a league softball game, the result will be accusations of cheating, perhaps a great argument, and probably the loss of all the fun involved in playing a game. If you want to play by yourself, you can do whatever you choose to do. But if you want to play with other people, you have to know what game it is you are playing, what the rules are, and you have to play by the rules of that game.

Chapter Two: Why Use the Label "Game"?

Just as checkers and chess are two different games, so science and religion are two different games with different rules. You cannot play either checkers or chess without first agreeing on the rules of the game, and likewise you cannot conduct good study in either the field of science or the field of religion unless you understand the rules of the game, that is, the rules which govern that particular field of inquiry. But I believe that disregard for the rules is exactly what many people do, and this is why there is so much tension connected to the relationship between science and religion. Some people try to bring the rules of religion into the field of science, and other people object to this confusion. Some people try to bring the rules of science into the field of religion, and this confusion results in objections by still others. In the fields of science and religion, it is necessary to know which rules you are using, and to abide by them. If you don't follow the rules of the game, you cannot play the game with others.

You are most likely wondering about the appropriateness of using the label "game" for the study of science or the study of religion. We do not usually think of the fields of science and religion as games. Games are artificial activities that we do for recreation. Science and religion are not recreation, but serious enterprises that seek to determine truth. Games follow artificial rules that we humans create, because they are not about reality. We make up rules for a game in order to limit the world in which the game is conducted.

But this kind of distinction between games and the study of reality is incorrect. Science and religion are both like games for two reasons. First, neither of them operates with total knowledge about the universe; that is, both of them are limited in their comprehension of the world, just as the playing space of a game is limited. Second, both of them operate according to rules that humans have established. It is because of these two facts that both of these fields, science and religion, operate just like recreational games. I think that it is helpful to bring the label of "game" into the discussion in order to remind ourselves of the limits that apply to both of these serious fields.

Consider the concept of war games which are often played by military forces. It is common for army officers to devise practice situations for military maneuvers, and to structure those practices as a contest between different groups of soldiers. In a war game, no one is actually killed by shooting or bombing. There are rules set out to determine who wins a particular part of the activity without actually having to kill the opponents. These rules involve who brings the superior force to bear, or who "gets the jump" on the opponent with a surprise maneuver. But despite these artificial rules, the conduct of a war game is still a serious matter, since it often reveals what would happen in the case of a real battle. The conduct of strategy and maneuvering in a war game is essentially the same as in a real battle. The fact that these military exercises are artificial and parallel to recreational games does not mean that they are not serious, and they are able to reveal the truth about the readiness of a battle group.

This analogy is also helpful because it is equally true that in the fields of science and religion no one is supposed to be hurt or killed. Both of these academic pursuits are supposed to be conducted in a manner that respects the safety and dignity of all the participants. Both of these academic fields understand that ultimately they are trying to comprehend very powerful forces that can be quite dangerous, in fact truly deadly, to human beings. In the field of science we seek to understand the forces of gravity, chemistry, heat, electricity, lightning, and thermonuclear reactions. In the field of religion we seek to better understand

God,[1] who in Christianity and many other religions is presumed by religious believers to control life and death, and who is usually considered the creator of natural forces and the one who will undo nature when he brings an end to this world. But just as military leaders conduct a war game in order to be better prepared for an actual war, so today scientists and theologians practice games, that is, conduct scholarly research in both fields in order that human beings may safely interact with the fantastic powers that they investigate. The research in both fields is carried out in a structured and somewhat artificial manner that preserves the safety of all human beings.

In the history of the field of science, there have been times when scholars have used human beings as the subjects of experiments in ways that disregarded the health and dignity of those humans. Humans have been subjected to diseases, or surgeries, or to various types of suffering in order to learn more about a disease, or to learn more about the functioning of the human body.[2] Today those kinds of experiments are considered unethical, and are not permitted. A similar kind of abuse of human beings has happened in the history of religion. The Christian Bible teaches that in the religion of Christianity "our struggle is not against flesh and blood, but against ... the spiritual

---

[1] It is common in the religions of Christianity, Judaism, and Islam to use the convention of a capital *G* when referring to the specific deity revered by that religion. The word "god" without a capital *G* will be used when the discussion applies not to a specific god or a specific religion, such as the God of the Christian Bible, but to any god in any religion.

[2] Examples include the Tuskegee Syphilis Experiment of 1932-1972, the many experiments with radiation carried out by the US military after 1945, and the experiments on prisoners by the Nazis during WWII. A beginning overview of these topics may be found on Wikipedia at "Unethical human experimentation in the United States" and "Nazi human experimentation."

forces of evil in the heavenly realms" that try to tempt and mislead human beings (Ephesians 6:12). Christians are instructed to "love your enemies and pray for those who persecute you" (Matthew 5:44), and to do good to those who oppose them (Luke 6:27, Galatians 6:10, 1 Timothy 6:18). This means that in the Christian religion the pursuit of spiritual truth should be conducted without aiming to harm other humans. The Christian apostle Paul wrote that "the goal of this command [to proclaim sound doctrine] is love, which comes from a pure heart and a good conscience and a sincere faith" (1 Timothy 1:5). However, in the history of the Christian church there have been many times when the pursuit of good academic doctrine has resulted in the harming of humans. People have been put to death, they have been whipped or beaten, they have been stripped of property and position, and they have been insulted, reviled, and vilified for the sake of academic disagreements. The Muslim religion holds a similar view calling for the respect of humans, and unfortunately Muslim history also has some sad examples of the religious followers not living up to that view. You only need to contrast the bombings and killings in Iraq today, carried out in the name of Sunni versus Shi'ite differences, with the behavior of Muhammad when he finally captured Mecca in AD 632—when people were astounded then at how kind he was to his enemies.

This is why it is helpful to think of both academic fields, the search for truth in science and the search for truth in religion, as games. Modern scholars have devised rules that these academic pursuits must follow, and they have limited the playing fields in certain ways so that the pursuit of academic truth will serve, rather than harm, human society.

As mentioned earlier, neither religion nor science operates from a position of knowing everything. Both are limited in their understanding of ultimate truth, and both need to respect that limitation.

Take religion first: one of the major truths that most religions offer is that humans do not and cannot know or understand everything that God knows. If humans could know everything that God knows, then humans could become the equal of God. If we use the Christian religion as an example, this idea that people can know what God knows is recognized to be one of the most fundamental flaws in human thinking. This was the original temptation used by the Serpent on the first woman Eve (Genesis 3), and it was part of the temptation in the primeval Tower of Babel story (Genesis 11). In the book of Job (Job 38-42), God's response to Job's questions is not to explain why God allowed Job to suffer, but to remind Job that he cannot understand how God created and runs the many facets of the natural world, so how could Job even attempt to fully understand God's ultimate control of the moral world? And in the Christian New Testament, when the apostle Paul discusses the mystery of predestination (Romans 9-11), he does not actually explain it, but merely asserts the truth of predestination, and then indicates that it is at this point that humans need to draw the curtain on how much they are entitled to understand about God's ultimate decisions. Religion is not about knowing all things, but about knowing those things that God has revealed. More will be said about this in Chapter Four.

In the same way the field of modern natural science does not and cannot attempt to know and explain all things. At a fundamental level, in modern physics the Uncertainty Principle and the formulation of quantum mechanics state that we cannot know every aspect of the physical situation of any phenomenon. In some ways modern science is almost the opposite of the knowing of all things. Modern science progresses by taking things apart and examining the details. The scientist limits an experiment by controlling as many variables and factors as possible, so that he can examine the effect of changing only one thing at a time. Once you have learned as much as possible about all the details, you still try to put it all together to look at the total picture. But that does not mean you understand the total picture, it only means that you are ready to examine some of the bigger features where all the little details interact. For example, no matter how many tests you perform in a wind tunnel, there is still the drama of the first flight of a new aircraft. In biochemistry scientists can only dream of the day when they might be able to devise a new medicine and then use a computer to predict exactly how it will work when used on human beings. The more we know about the science of meteorology, the more we have learned to stop trying to predict the weather perfectly; instead we have learned to provide only a range of probabilities as to whether it will rain on the coming weekend. Modern science is based on the idea that we do not know everything, and that we must continue experimenting to try to learn more about each part of every natural process. One of the major principles of modern science is that the scientist must always carefully determine what it is that he can know and what he cannot know: what does a particular experiment reveal, and what does it not reveal.

    Though some people argue that all of our existence and experience in life can be reduced to a scientific explanation of the functioning of matter and energy, most religious people (who make up over 85% of the world's current population) do not agree with this. Most people think that modern science cannot account for all facets of reality. For example, the studies of the processes of mating among animals, and the mysteries of DNA and the replication of living things, do not provide adequate grounds for instruction on the morality of human dating, marriage, or divorce. There is much more involved in the process of human reproduction than just the chemistry with the pheromones and hormones: at the least, psychology, sociology, and religious beliefs play various roles in marriage decisions.

The concept of values enters into any discussion of what we should do, and science is not about the examination of values. Scientists use values to decide about research, but science as its own discipline cannot define or create a system of values. It is very seldom that someone has the time and resources to try every possible experiment in connection with some topic. Scientists have to make decisions about which experiments to fund and conduct. In doing so they are guided by their estimation of the value of the time and cost of any activity versus other concerns. The analysis of values moves from the topic of natural science to what we call the humanities, the study of things pertaining to human thinking and experience. In the study of the humanities we include history, philosophy, ethics, drama and literature, linguistics, religion, all the arts, psychology, sociology, anthropology, and political science. Topics such as law, business, and education are also part of the study of human life, though we often separate those topics from the humanities into their own fields because of their specific focus on an aspect of human life. In a similar way engineering is often separated from the study of natural science because, though it requires much knowledge of science, engineering focuses on various applications to human life. Engineering requires decisions about the value for humans of various constructions, but it is usually not considered to be part of the humanities; however the increasing attention to medical engineering may require some adjustment to traditional thinking.

The discussion of values includes more than the review of ethics. It includes the discussion of cost, and of beauty. Although scientific information can often help in making a better analysis of the relative cost of some venture, scientific thinking by itself cannot decide the relative value of different choices, nor can it analyze the ethics or beauty of some activity or construction. Science interacts with art in many ways, and in the field of science, some aspects are even labeled more "beautiful"[3] than others—such as certain formulas in physics or mathematics. But while scientific examination can help us determine some of the factors that lead us to consider something beautiful, science by itself cannot define what beauty is. The field of architecture is a good example of something that requires a solid knowledge of science and engineering, but which requires something beyond this. A good architect must balance function and cost with that thing we call "good design," that sense of how shape and space affect human perception.

Neither science nor religion claim that humans know everything, or even that they can ever achieve the point of knowing everything. Therefore, when someone sets out to "play" in either the field of science or the field of religion, the process is parallel to playing a recreational game, in which there is a defined (limited) playing field and rules for the conduct of the game. Humans devise a set of rules for how to structure the process of discovery in either of these two fields, and they learn to play by these rules. More will be said about the rules for science and the rules for religion in Chapters Three and Five.

There is another way of looking at the concept of games to help understand the parallel structures of science and religion. In a track and field meet, there are many different kinds of competition. There are races, jumps, and throws. Each of the different competitions has its own rules. It does not matter how fast you throw the javelin, only how far. It does not matter how fast you approach the high jump, only how high you jump. You

---

3   Terms such as elegant, neat, concise, useful, pleasing, attractive, handy, suitable, tidy, simple, or nifty may be used by different writers, but all of them point to a judgment of the value or the mental aesthetic effect of the item.

cannot use the rules for one contest to judge another. In the relay races, it is essential that you do not drop the baton. In the pole vault it is essential that you do let go of the pole at the proper time. In the short races you have to stay in your lane. In the 10 kilometer race the lane markers are irrelevant, and no one has to worry about crossing them. In the steeplechase you must turn off the track at one point to make the water jump. If you try to use the same shortcut during the 10K, you will be disqualified. You only earn points in each contest if you follow the rules for that contest. But the track and field meet is decided not by one contest, but by totaling up the points from all the contests.

In a similar way, in the great track meet of life, we humans compete in several different events. We seek truth through both science and religion. But you have to follow the rules for each game independently, or you can't score any points. It can be very tempting to find a compromise between science and religion that brings them both together into one scheme. But if that involves violating the rules of either game, it is not valid. Just as in track and field, in life the rules of science must be followed, and the results totaled, and the rules of religion must be followed, and the results totaled. Then you add together the results from each endeavor. If you have transgressed the rules of either game during the track meet of life, your results do not count. In track and field, it may be that you ran down the runway for the long jump with the fastest speed, and there is a relationship between your speed and the distance you jump. But the final result is not scored for your speed, but for the distance of your jump. You get no points in the long jump just for speed. In a similar way, you get no points for applying the concepts of religion in the game of science. Perhaps your laboratory experienced the most kind and ethical treatment of all employees in the entire scientific community. There is a relationship between how one treats employees and the ultimate outcome of the cooperative endeavor. But the final result by which a laboratory is judged is the scientific progress, not the social experience. If you become obsessed with the pursuit of a harmony of ideas, such that you cross the rules of science and the rules of religion, then your exertion does not count and it does not win you any points with other people who are engaged in either of the separate searches for truth.

It may be useful to consider one more example of how the academic fields of science and religion are both intentionally artificially limited, and thus are parallel to recreational games. This last example is the field of criminal justice. You might think that the criminal justice system is the antithesis of a recreational game, for the outcome can be very serious, often having a dramatic impact on the lives of the participants: fines are assessed, property changes hands, people go to jail, or even get sentenced to capital punishment. In other cases the custody of children can be switched from one parent to another relative, or even given to the state. But despite these very serious results, the conduct of the criminal justice system is parallel to the conduct of a recreational game. We have set up rules that all the participants must follow. If you break the rules, you cannot win the game: the criminal justice system makes decisions against you. If you fail to show up in court at the proper time, you might well lose your court case, even if you had all the evidence and facts on your side. The court does not simply determine and award justice; rather, it determines who has done the best play according to the rules, and determines the winner in the light of that. We assume that if we have managed to set up the rules of the criminal justice system as best we can, then over the long run the results will tend to be as close to justice as we can get.

Take the case where the police have conducted an illegitimate search and found evidence against an accused person; say they have found some illegal drugs in a person's pockets during a search that they had no proper reason to conduct. In this case we have established the rule that injustice carried out by the police cannot, in the long run, work toward the establishment of justice in society. So if the police break the rules by conducting an illegal search, the evidence discovered in that search cannot be used in a trial against the accused. If the police obtain a confession from the accused through some illegal coercion, that confession is ruled out of bounds: it cannot be used for any legal purpose at a trial. These rules sometimes produce the result that the court declares as not guilty a person who we know is, in fact, guilty. We know the person did have illegal drugs in his pockets. The police know it, the judge knows it, and the accused person knows that the others know this. But without other evidence and with this illegally obtained evidence ruled out of play, the accused person is not convicted, and he wins the case. The state contributes a significant outlay of money for the arrest and trial of the accused, but in the end the person is set free.

All of this is very serious. The criminal justice system does not establish justice: it establishes the closest thing to justice for those who best follow the rules society has established. We sometimes read that police and attorneys fume that the criminal justice system is just a game that criminals know how to play. This is correct—there is a real element of truth in such a statement. We have deliberately structured our criminal justice system in this way. The reason we have created these careful rules and we insist on following them is that we humans do not and cannot know everything, including all the facts about human interaction and criminal conduct. So we have painstakingly devised these rules about the collection of evidence and the conduct of the entire process of accusation, investigation, and trial in order to insure that injustice does not happen through our limited knowledge. You have to play by the rules to win in court. You may know that someone has harmed you in some way, but if you cannot prove it in court, you are not entitled to any compensation or any type of revenge in our system of criminal justice. We have intentionally created a system of justice that functions in an artificial way, just like a recreational game, in order to ensure that the criminal justice process suffers from the least abuse.

The Distinction Between Natural and Supernatural

Now that I have explained why the operation/conduct of both science and religion can be thought of as playing games, there is one other topic that deserves attention at this point. If I am going to separate the search for truth into two different games, the game of science and the game of religion, what is the basis for making a distinction? How do you know which game you should play at any point? In other words, how do you know which rules you should follow in conducting your inquiry? The distinction that I make in this book is between investigating that which is natural, and that which is supernatural. The rules of science are used for the investigation of natural phenomena, while the rules of religion are used for the investigation of supernatural phenomena. So how does one distinguish between the natural and the supernatural?

First, you must grant that a distinction can and should be made. If for any reason you deny that there is any supernatural force that acts within the world, then everything that happens can only be considered natural. There are many people today who try to hold to this view, the denial of anything supernatural. But one problem with this view is that it provides no basis for determining significance. This brings back the matter of values, as mentioned above. Natural science procedures by themselves cannot assign value, significance, or meaning to any activity. A fallen rock is just that: a fallen rock, and nothing more. It requires human thinking to assign significance to any event in nature, or in history. When events happen, humans tend to ask the question "why." Why did this country win the last war? Why did I obtain this particular job? Why did this special person come into my life? Why did one person survive an accident, and another person die? Many people believe that in finding some meaning in answers to such questions, they perceive a connection with the supernatural powers. The events are correlated with their sets of values, and thus the events fit into some significance with regard to purposes that are greater than the mere events, and thus are "beyond" the natural events. This concept of purpose that is "beyond" the merely natural invokes the existence of the supernatural. Most religious people are so used to thinking in this manner that it is very hard for them to try to reason within a framework that excludes such connections. Such religious people often do not realize that it requires great courage and great self-discipline to maintain a denial of any meaning pertaining to the events of life. The assertion and the attempt to hold to the assertion that there is no meaning in life is a difficult and often frightening thing to carry out. If there is no supernatural, then there is only the natural. But in fact most scientists believe there is some kind of meaning, some kind of beauty, some kind of value in life. Because we do not yet (and never will) know everything about the universe, this sense of meaning and the search for meaning keep pointing toward the existence of the supernatural.

If the one extreme is to deny the existence of the supernatural, the other extreme is to assert the total role of the supernatural in such a way as to deny the existence of the merely natural. Religious people believe that God is responsible for all phenomena, that God is the creator, preserver, and controller of all phenomena, both natural and supernatural. The temptation then is to conclude that the game of religion always overrides the game of science, and the rules of religion should determine the ultimate truths of science. But such logic erases the question of the distinction between the natural and the supernatural instead of answering how the distinction is to be made. If the rules of the game of religion always apply and always supersede the rules of the game of science, then there is no separate game of science. There is no distinction left between what is natural and what is supernatural, and that means in effect that there is no such thing as "supernatural" because everything belongs to the domain of nature, and the concept of "God" is reduced to the functioning of nature. In this case every natural thing that happens is perceived to carry "supernatural" meaning, and there is no such thing as random chance. If you get sick, God wills it; if you get fired from your job, God wills it; if you stub your toe, God wills it. Despite our human attempt to detect any cause-and-effect relationship between our actions—the decisions and actions we carry out—and our fate, this extreme view overrides such logic, and concludes that everything that happens must be the will of God. The will of God cannot be thwarted; therefore we must simply accept our fate. Then there is no place and no purpose for science. Indeed, in such a view when everything is considered supernatural, science can be thought of as trying to contest the will of God!

In between these two extreme views is the view that makes a distinction between natural action within the world and supernatural action intervening in the world. Humans believe that there are laws of cause-and-effect that govern how most things work in the world. But sometimes something seems to override the known rules, and we acknowledge that mysterious intervention. If you grant the existence of the supernatural for the purpose of discussion, then you have to define how you will distinguish between natural events and supernatural events. In other words, for the religious person, how do you know when it is proper to analyze something as natural, and apply the rules of the game of science, and when it is proper to analyze something as supernatural, and apply the rules of the game of religion? For the religious person, this distinction must be founded in something that has warrant from God. **The distinction I offer is simple: when God chooses to act in regular and predictable ways, we are to recognize that behavior and we are to investigate God's design of the natural order. When God chooses to act in unpredictable ways, we are to investigate God's behavior by methods that include contact with the supernatural.**

We humans have learned to create the game of science for the study of things that are natural. We have learned in different ways through history that though God may be the creator of the world, the created world runs according to natural laws: it follows certain principles of cause and effect that we can perceive, understand, and predict. Some of this understanding of the existence of natural laws has come about from painful trial and error, such as the realization that susceptibility to disease is a matter of nature and not a sign of divine favor or disfavor. Some of this understanding apparently belongs to humans as far back in history as our knowledge can go. Whatever it means to be a human being, part of that includes having a mind that is able to perceive cause-and-effect relationships and make calculations that include such knowledge. Humans have been able to make a distinction between what is natural and what is not from the time of their creation, and they have slowly expanded their knowledge of natural laws and natural functioning throughout history.

The fact that the sun rises every morning to begin a new day is both regular and predictable. Before we understood the physics of the rotation of the Earth, we had no way to guarantee the prediction that the sun would rise the next day. But the obvious regularity of this phenomenon made it a pretty safe bet. The rotation of the seasons, the rotation of the stars and constellations, the pattern of crops and wild vegetation, and the migration of animals are all regular patterns that display God's creation of natural laws. Their regularity defines them as natural, as part of the designed and predictable process. Something potentially qualifies as supernatural only when it does not fit this regularity or predictability.

There are certain phenomena that ancient civilizations often grouped into the category of supernatural because of their unpredictability, but which nevertheless had aspects that suggested they were not simply supernatural. Among these we might list the striking of lightning, or the eruption of volcanoes, or the occurrence of earthquakes or tsunamis. In one sense it was and still is pretty hard to predict where lightning will strike. But it was obvious that lightning tended to come with a thunderstorm, and almost never struck out of a blue sky. This linkage suggests that, for all of the unknowns associated with lightning, it is not simply a supernatural phenomenon: it is part of nature, and tied in to the rules for rainfall. Volcanoes and earthquakes are unpredictable, but those people who live in certain areas learn quickly that there is a certain long-term regularity of their occurrence. Today we know about fault lines, magma channels, and tectonic plate boundaries. Our vastly greater knowledge makes clear that these phenomena fall into the category of the natural. Another example is the affliction of diseases. For centuries humans could not explain why some people contracted a disease and died, and others did not contract it, or contracted it and recovered. Thus prayers to the supernatural world were a part of the treatment of disease (and still are a meaningful part of the treatment today). But even before the rise of the modern science of microbiology, humans could determine certain aspects governing the probability of the contagion of different diseases. Thus the basis existed for the scientific investigation of disease, when the technology finally became available. Disease is not simply the random will of a capricious god, but it is something natural that follows natural laws. Experiencing a disease, or investigating a disease, does not involve the pleasure or displeasure of God. For the religious person, the matter of the regularity and predictability of all such phenomena require them to be classified as natural phenomena, and thus necessarily subject to human investigation. In Chapter Five I will talk more about revelation from the supernatural world. At this point it is adequate to point out that without some type of guiding revelation, natural phenomena cannot be taken as clear indicators of God's will. The fact that the sun will rise tomorrow does not prove anything about the existence or non-existence of God, or about his favor or disfavor toward certain individuals. Such events as eclipses, or the passing of comets, were often interpreted in the past as supernatural events, but today we know enough about their regularity to label them as natural. We have learned to distinguish between the natural and the supernatural: when God acts in regular and predictable ways, we investigate the laws of science; when the laws of science do not apply, then we raise the question of the supernatural—but we still need some revelation from the divine world to guide any further reasoning.

Chapter Three: The Game of Science and the Rules of Modern Science

A Brief Review of the Creation of Modern Natural Science

So how is the process of modern natural science like a game, and what are some of the rules that participants must know and follow? Let me begin to answer this by giving a brief summary of the development of this game.

The game that we know as the modern field of natural science began in Europe about five hundred years ago. Before the rise of European science, there had been very significant discoveries and inventions in ancient China, in India, and in the Middle East. Special credit must be given to the great advances of learning that happened in the flowering of the Muslim civilization after AD 750. The scholars in the Muslim civilization, headquartered at that time in ancient Persia (modern Iran), built upon the knowledge of the ancient world (including the more ancient knowledge of Egypt and Greece) in many ways. These Muslim scholars created the foundation for modern mathematics, astronomy, medicine, chemistry, banking, and other fields. The Muslim scholars also investigated the writings of the ancient Greeks, translated them into Arabic, and brought them into discussion in their civilization. Thus the Muslims incorporated into their development of scholarship all of the great early Greek work in mathematics, science, logic, and philosophy. The Europeans had no interest in any of this old Greek literature at that time, for Europe was in what is sometimes called the "dark ages." As much as modern Europeans might make claim to this heritage of the great scholarship of ancient Greece, it came to Europe through the work of Muslim scholarship, not directly through European scholarship.

But while this flowering of scholarship in the Muslim civilization was great, still something different happened later in Europe that produced the modern field of natural science. Modern natural science owes a great deal to its Muslim heritage. But the game of science as played by the Muslims in the 8th through the 15th centuries was not quite the game that developed later in Europe. There was a change in the rules and thus in the game of science as it evolved in Europe, and it is this later game that is the modern field of natural science. What was this change, and how did it come about? Scholars still struggle to explain why this change happened in Europe. I will be bold enough to suggest that the topic of this book, the distinction between science and religion, is a significant part of the reason behind the change.

Around AD 1100 the leaders of both church and society in western Europe chose to launch the Crusades (AD 1100-1300) to try to conquer the city of Jerusalem and the territory of Biblical Israel in order to take it away from Muslim control and bring it back into Christian control. The first Crusade was successful. But violence often begets violence, and the Muslim empire soon re-conquered Jerusalem. The later Crusades were not successful. However, these crusades brought many of the leading men and soldiers from western Europe to the Middle East, and exposed them to the much more advanced culture of the Islamic civilization. When they returned home to Europe, these Europeans sought to bring much of this advanced culture and technology with them. Many of them hired Muslim scholars to establish universities in their home countries, universities modeled after the Muslim universities. The influence of this heritage of the Muslim university cannot be overstated. Consider just two facts to illustrate this. One, the term "algebra" is Arabic for "the number," or the science of numbers, and it is the label for a significant part of the study of all mathematics today. Two, today western universities still conduct graduation ceremonies in which all the participants wear robes that many scholars argue are the heritage of the Arabian desert robes, as established in the Muslim academic tradition.

The founding of these European universities gave birth to a revival of scholarship in the European world. This rebirth of literacy and learning is called the *Renaissance* (French for rebirth) in European history. The Renaissance (14th-17th centuries) brought forth its own offspring, including both the European religious Reformation (16th century), and the Enlightenment (17th-18th centuries). Once significant portions of the European population resumed reading, they also renewed the reading of the Christian Bible, and that led to the Protestant Reformation, a movement to make corrections in Christian tradition to bring the faith back into alignment with the Bible. The Enlightenment was a new philosophical view to sit alongside, or perhaps replace, the traditional European religious view that had been dominant for centuries (more will be said about the Enlightenment below).

But the process of adopting this scholarship from Muslim culture involved a subtle but important shift in thinking in western Europe. The very title "university" points to the goal of seeking to bring together all the realms of knowledge: the goal was a "universal" picture of all truth and all information. The truths of religion and the truths of science were both part of this unified system of knowledge. This Muslim scholarship included a significant history of the discussion of the relationship between religious faith and non-religious reason. Muslim scholars were aware of the fact that sometimes religious reasoning and non-religious reasoning produced ideas that were in conflict. Significant effort had been devoted to harmonize these two processes of reasoning, and to describe the rules for how these two sources could be integrated. When the Christian Europeans borrowed all of this Muslim scholarship, they had to adapt the religious component away from Islam to Christianity. Along with that they had a more subtle issue to handle. As is true with most cultures and religions in history, Christian Europe believed that it alone had the true religion, that it was the one culture closest to God and having God's favor. But in borrowing this Muslim scholarship they had to adjust to the idea that even though they thought they were the people most beloved of God on Earth, it was obvious that God had allowed a much higher and better culture to develop in another land and religion. If Christian Europeans were indeed God's favored people, how could this have happened? The resolution of this theological problem required the assumption that there was a difference between having the fullness of religious revelation and knowledge, and having the best of technical or worldly scholarship. This assumption prepared the way for the separation of the games of religion and science.

    The Muslim world of scholarship already recognized the distinction between the truths of the natural world and the truths of religion. Their holy book, the Quran (or Koran), encouraged learning about the natural world as God's gift to humans. Thus they had developed great learning about the natural world. But the Muslim world did not separate the truths of science from the truths of religion; they understood the distinction, but they did not make a separation. The European Christians were forced toward making a separation in order to maintain the validity of their religious reasoning. The Muslim religious emphasis on the unity of all things under God's almighty power allowed them to maintain a conceptual unity of all truth, such that science and religion were completely compatible. However, in building on the work of the ancient Greek philosophers, the Muslims also carried forward the philosophical discussion of what constitutes truth, and how do we perceive, detect, or know truth. This involves the academic topic of philosophy.

This book discusses the difference between the two games of religion and science. At this point some attention must also be given to a third game, the game of philosophy. **Philosophy can be defined as the study of human reasoning. The game of science looks at the truths of the natural world, and the game of religion looks at the truths of God, while the game of philosophy looks at the truths of the human reasoning process.** (This will be Rule Two in the summary at the end of this chapter.) Obviously human reasoning is used in all three of these games. Perhaps it might help to understand this if I describe a parallel situation.

When you work in the field of engineering to build a dam, you make much use of mathematics. When you work in the field of political science, such as assessing the opinions of population groups as to whether or not to build a dam, you also make much use of mathematics. But it is also possible to study mathematics for itself, without reference to the building of dams or the assessment of opinions. Mathematics is used in the field of engineering, and mathematics is used in the field of political science, and is also studied as a field in itself. Just so, we use human reasoning in the study of science, and in the study of religion, and it is possible to simply study the process of human reasoning itself. Thus far this book has used the word "truth" without giving any attention to its definition. Philosophy is the branch of human scholarship where people wrestle with the concept and definition of truth. What is knowledge? How do we know what we know? What is the relationship between reality and our human experience? For example, how is our perception of light connected to reality? Does our vision represent reality, or only an illusion? We can "see" and sense light waves, but we cannot sense radio waves; therefore are radio waves real? In what sense is our logical reasoning valid? Did that stage magician actually make the woman disappear, or was that merely an illusion? How does good logic function? If my dog attacked and bit Sam, is it logical to conclude that my dog will now attack and bite me also? If I know from watching cowboy movies that the good guy always wears a white hat, and I come across a strange dog that has white fur, may I deduce from the color that this is a good dog which will not hurt me? These kinds of questions involve not simply science or religion, but the larger realm of thinking that we call philosophy.

In the study of mathematics we determine certain things to be "true" in mathematics. We then use those "truths" of mathematics when performing calculations in either engineering or political science. In the study of philosophy we determine certain things to be "true" in logic and reasoning. We then use those "truths" of logical philosophy when working in other areas of life, that is, in other games. In the case of one of the above examples, we would reason that the color of a man's hat or a dog's fur is not adequate information to determine its potential danger. The characteristic of white color does not necessarily correlate with the characteristic of non-malicious behavior.

The matter of the relationship between the fields of science and religion is often discussed in different phrasing: the relationship between *faith and reason* as sources of truth. But this phrasing is a different orientation to the discussion. The phrasing of faith versus reason shifts us out of the game of science into the game of philosophy, as these games are being defined here. This discussion of faith versus reason does not focus on the facts of natural science or on the objective content of religion, but on the question of how do we understand what is truth coming from different sources. Is human thinking—reasoning—a valid way to obtain truth? Is placing faith in some religious authority a valid way to obtain truth? This discussion about the validity of reason is not the same as the discussion about the knowledge of science or the knowledge of religion. And this distinction of games points back to the importance of knowing which game you are playing, and what the proper rules are for that particular game. You cannot begin playing checkers, and suddenly shift to using the rules of chess. You cannot begin playing the game of science, and suddenly shift to using the rules of the game of philosophy. You can borrow the truths of logic from philosophy into science, but that does not make the rules of the game of science the same as the rules of the game of philosophy. Yet this error is exactly what is commonly done.

    The ancient Greek philosophers were not simply interested in natural science. They were also interested in the study of art, and virtue, and human reasoning. They sought to bring all these fields together into a logical scheme that they believed would explain everything. Muslim scholarship pursued this same goal. Jewish scholarship within the Muslim world was able to develop its own great tradition, and it also pursued this goal. And when the Europeans borrowed all of this tradition of scholarship from the Muslim world and joined in the conversation, they also pursued this goal. Thus the search for the understanding of truth was almost always a mixture of the study of natural science, spiritual questions, and philosophical questions. In other words, at some level the games of science, religion, and philosophy were all combined and played together. Return to the label that is used today for an institution of higher education: it is called a "university." A university education is intended to provide the student with an education that looks at and brings together all the information from all the different types of research so as to form a complete view of how all truth fits together. This education was thought to form one coherent view of everything, all truth fitting together: it was universal in scope. The ironic fact today is that at modern universities we keep subdividing and separating the different fields of study into different disciplines, because we realize that the different fields are not the same. They require different methods of research, and they produce different kinds of data. Psychology is not the same as sociology, theology is not the same as philosophy, chemistry is not the same as physics, and anthropology is not the same as biology. A good compromise in a business deal is not necessarily the same as a good compromise in a political deal. Today a good university education provides the student with a sharpened knowledge of the differences in the various facets of life, and that one must be aware that things do not always fit together into one unified scheme of understanding.

Today we often overlook the history of this mixture of games when we review the history of natural science. We hail the significant advances made in the natural sciences and mathematics by people such as Rene Descartes, Galileo Galilei, Isaac Newton, Benjamin Franklin, or Charles Darwin, while we ignore the fact that as people of their times and cultures they were also very involved in linking their knowledge of natural science with their theories of spiritual and moral truth. This distorts our understanding of history; it also fosters an inaccurate understanding of these particular humans in which we view them as if they were twenty-first century minds living among seventeenth century contemporaries. We ignore the spiritual musings of Newton; we do not ponder the effect of Franklin's membership in the Masons; we ignore the family history of Darwin and the effect this had upon how he chose to write about his scientific research. The fact that we can strip apart the contributions of these people in the field of science from their related speculations in other fields such as theology or political philosophy demonstrates the importance of keeping the games separate, and knowing which game you are examining.

Here is another analogy. When you build a bridge, you use what you have learned from the study of mathematics, physics, and chemistry. From physics it is important to understand how the forces of gravity, wind, water and river current, and traffic load weight will act upon all the parts of the bridge. From chemistry it is important to understand how the different materials used in the many parts of the bridge will interact: which parts can withstand being wet without weakening, which parts will interact with rain and other chemicals in the atmosphere, and how will they affect each other. From mathematics we take the rules for performing the calculations to adjust the size and strength of all the pieces of the new bridge. We use all these different sciences to help design and build a bridge. But it is important to use each properly: we cannot use mathematics to answer a chemistry question, or chemistry to answer a physics question. In the same way, when we seek to learn all the truth of reality, we use the tools of natural science, religion, and philosophy. But it is important to know which tools we are using at each point. We cannot use philosophy to solve a science question, or science to solve a religious question. We have to be aware of which field we are engaging. We have to know which game we are playing, and what the relevant rules are for that game.

Muslim scholarship during the Middle Ages had developed natural science to a new level. But the Muslims had no reason to consider any incompatibility between the truth of science and the truth of religion. So they were willing to harmonize these fields when they felt it appropriate. When they considered any tension between these fields, they moved into the game of philosophy, and discussed the relationship of faith and reason. The prevailing idea was that all truth came from God, and therefore they tried to harmonize all truth as if they could see it from the viewpoint of God. When the European world of scholarship adopted all of the Muslim scholarship resulting in the Renaissance, the Europeans followed in that same tradition of trying to conceive of all truth working in harmony as seen from the viewpoint of God. The discussion of the relationship between faith and reason was a part of the study of philosophy. Philosophy, theology, and natural science were all mixed together in the attempt to define truth as viewed from the divine standpoint. But along with that tradition, something else, something new, happened in Europe. I suggest that because of the religious crisis felt by the European Christians (the problem they felt in believing they had the best religious relationship with God but not the most blessed worldly situation), the philosophical linkage between natural science and religion was severed, such that the Europeans became able to "play the game" of natural science without being limited by any compunction to harmonize the results of scientific observation with their theology. Though the European scholars continued to interplay the games of science, religion, and philosophy, they also felt free to play one at a time, and the game of science slowly developed in a new way in the context of this different situation. I mentioned earlier that the borrowing of Muslim scholarship which resulted from the Crusades led to the Renaissance, which in turn led to two great changes in Europe: the Protestant theological Reformation, and the Enlightenment. The Enlightenment can be understood as this separation of viewpoints (or in terms of this book, the distinction of different games). After centuries of the "dark ages" in Europe when all academic and philosophical thinking was dominated by religion, this rebirth of scholarship in Europe proceeded along a path that held that there were other ways of viewing things than merely the Christian theological way. This freedom to examine and use other ideas allowed the development of many new movements in philosophy, politics, theology, and science.

In the game or field of natural science, we can trace a fundamental adjustment back to a French scholar named Rene Descartes (pronounced *de-KART*). Descartes lived from 1596-1650. This period is after the Reformation and after the discovery of the American continents, but before the age of Napoleon and the revolutions in America and in France. It was the time of the setting of the novel titled *The Three Musketeers*. Descartes was a contemporary of Galileo, who invented the telescope; but Galileo lived in Italy rather than France. As far as religion goes, Descartes was a faithful French Roman Catholic all his life. He was a great scholar and mathematician. Among other things, he produced significant work in graphical linear algebra: the Cartesian coordinate system is named after Descartes for his development of the system, wherein you can locate a point on a graph by assigning it x and y coordinate values. But Descartes is also usually recognized as the foundational figure in modern European or western philosophy.

Descartes wrote about human emotions, rationalism, and the search for truth. Descartes is credited with laying the foundation for modern philosophy by changing the orientation of philosophical research. Instead of trying to determine what God had laid out as truth, Descartes shifted the focus to what he as a human could determine to be true by his own efforts. We can describe this as separating the games of religion and philosophy, and letting each develop their separate rules.

But Descartes' work laid the foundation for the modern game of natural science, though this was not appreciated in his time and was not thought of in such words. During the lifetime of Descartes (in the period following the Protestant Reformation), religion was up for debate. Roman Catholics had one Christian theological system, German Lutherans had another, Swiss and German Calvinists had a third system, the Anabaptist movement in Europe had another system, and the Church of England was experimenting with yet another system by going in a different direction for the relationship of church and state. The Eastern Orthodox Christian tradition refused to agree with either the Roman Catholic system of theology or the new Protestant theologies. The English Pilgrims would emigrate from England to

Holland and then to America during Descartes' time, seeking a place to practice their own view of religion. During this period, if you sought to determine what was truth, theology was only partially helpful, for you had to select from a number of different opinions as to which theology was correct. You might favor one opinion over another, but you could never escape the fact that you were making a choice among options. So how could a person determine for certain what was true? Descartes developed a philosophical system that built on human observation and logic.

Descartes is perhaps most well known for his statement "I think, therefore I am." This represented his resolution of a profound philosophical question: how could Descartes know if he was real, or if he was just a dream in the mind of God? Descartes chose to operate on the presumption that God would not lie to him. Therefore, if Descartes was able to exercise skepticism and question the matter of his existence, that meant that his inner self that was doing this reasoning was distinct from the reasoning. If not, God was producing a dream in which Descartes was being fooled into thinking that he was thinking. That would be a lie. Thus, the fact that Descartes was able to think and question meant that he had some foundational essence, some kind of mental presence in the universe. And he could proceed from that point.

But in developing this philosophical system, Descartes made a fundamental decision that greatly affects the game of science today. Descartes wisely concluded that he could not think or know the thoughts of God. In the area of the divine, he was limited to what God chose to reveal. And even God's revelation had to be interpreted by his human mind. So Descartes simply said that the realm of God was beyond him, and not the subject of his research. Descartes separated the truth that was above from the truth that was below. Today we would say he distinguished between the supernatural and the natural. For Descartes, the subject of human research was the natural: what could humans observe and determine from their observations? In Descartes' time philosophy and science were not as clearly distinguished as we have them today. So this careful distinction on Descartes' part was a philosophical decision, but it also affected the entire field of natural science. In effect, with this distinction Descartes separated the game of religion or theology from the games of philosophy and science, and laid the foundation for the separation of the game of science from the game of philosophy.

This distinction is fundamental to the game of modern natural science: **Science is only concerned with the natural, with what human beings can sense, observe, and somehow measure within the known universe. Science is not concerned with the supernatural. The supernatural is, by definition, outside the bounds of science.** (This will be Rule One in the summary at the end of this chapter.) The supernatural is beyond—"above" (Latin "*super*")—the natural. Science is concerned solely with what we humans can sense and observe, and agree upon. When you play the game of science, you cannot interject any information that comes from God. This means you cannot bring any revelation into a scientific discussion, not from the Bible or the Quran, nor can you draw any conclusion about the will of God from what you observe in science. If you build a tall building, and it is struck by lightning and destroyed, you cannot say that God desired to destroy the building. If you rebuild it and it is struck again by lightning, you still cannot say anything about God's will. If this happens three, four, or even five times, you still cannot explain this event scientifically as the will of God. In the game of religion, which I will examine in the next chapters, you must deal with the fact that in some sense everything that happens is the will of almighty God. But then you have to use the rules of the game of religion to decide what part of God's will any given event represents, for sometimes God allows bad things to happen in conflict with his greater will, such as allowing good people to suffer illness. In the game of science, you cannot bring this discussion into play. Certainly God allowed, or caused, the lightning to strike your building. But you cannot say why. If you stand up suddenly in your canoe, and it rolls over, you do not blame that dunking on God: you blame it on your poor seamanship. God just allows certain "natural" laws to play out. So if you want to know why lightning keeps striking your tall building, you are not allowed to consider any part of God or religion in that research. You must seek a natural reason for these lightning strikes.

For example, if you are a Protestant Baptist, and your tall building was struck by lightning, but the building of your Roman Catholic neighbors was not struck, you might be tempted to draw the conclusion that God is angry at your tall building because you are a Baptist, and that God favors Roman Catholicism. But all of your Protestant friends would protest your conclusion and reject it as unscientific.

Once again we can draw a very helpful parallel from the game of checkers. When you are playing checkers, all the pieces of both sides move on the black squares. Half of the squares on the playing board are colored red, but no piece ever sets on them. The game of checkers is carried out entirely on the black squares. As far as the game of checkers is concerned, the red squares do not exist. You could cut them out and throw them away, and it would not affect the game, as long as you kept the positioning of the black squares unchanged. In the game of checkers, the rules limit the activity to only using the black squares. Consider the black squares to be natural, and the red squares to represent the supernatural. In the game of science, you can only maneuver and reason with the natural, and you cannot ever invoke or reason with the supernatural. In the game of science, as it has been defined and developed from its foundation in modern Europe, the supernatural is out of bounds. Like the red squares in the game of checkers, the supernatural might as well not exist when you are playing the game of science. You are never allowed to move into that realm, or bring anything from that realm into play in thinking about science. If you do, you break the rules, and your reasoning is rejected by all other players.

Let me repeat: when you are playing the game of science, you are limited to the study of the natural world. You cannot in any way comment on anything pertaining to the supernatural. That is off limits, by the very definition of the game of science, as traced back to the fundamental change in orientation begun with Descartes. One of the rules of science that will be discussed below is that you have to be engaged in conversation with other scientists around the world. As part of that conversation, scientists agree not to bring their theological or philosophical speculations into the discussion. As far as joining together in the game of science, they agree on this fundamental principle to consider only what is natural. What are the natural objects and processes that we can sense, observe, and measure? The supernatural is out of bounds—the red squares do not exist. Of course, that does not mean that on a checkerboard the red squares do not really exist—they are right there. But they are not in play. And nothing about this foundational rule says that the supernatural does not exist. It only says that you cannot use it in any way when you are playing the international game of science with all your colleagues. You are free to go off and write any kind of theist or atheist philosophical book on your own time, and perhaps it will interest some of your colleagues. But that is outside the game of science.

However, this other kind of speculation that mixes science and religion is what has often been done in the history of the world. Many people find it attractive because, in their search for truth, they think that people who do well in the game of science must be very smart and so perhaps also know best the truths of the game of religion, or the game of philosophy. This interplay of the different games has existed all through history. Thus ancient Greek philosophy speculated on the linkage between beauty and mathematics. Aristotle tried to define items in existence as having both substance (their actual essence) and accidents (their apparent characteristics, such as color, temperature, weight, density, etc.), bringing together philosophy and science. This allowed for speculation that one could somehow change the characteristics of one thing into those of another thing, thus transforming the thing for practical purposes. Modern science rejects this view of Aristotle, by providing a different definition of what constitutes an object in existence. Thus you cannot change lead into gold. Consider some examples of this mixing of the games by great scholars in history. Isaac Newton's quest to discover and describe the fundamental principles that governed action in the universe influenced him to also reject the orthodox Christian view of the Trinitarian God and speculate upon a mystical spiritual unitarian scheme (note the linkage in trying to find a single unified view of truth). Charles Darwin felt that a mechanistic view of biological evolution solved the question of why God would allow his daughter to become sick and die, without any apparent divine concern for the child. Karl Marx believed that social evolutionary forces slowly defined human history, and that any talk of God or the supernatural was merely a tool used by some humans to control others. The Soviet cosmonauts rode their spacecraft up into the heavens, the realm of the gods, and came back to report that they did not find any god up there, thus he must not exist. Carl Sagan wrote about how the success of science in exposing the vast size of the universe suggests that all human religion is merely limited primitive speculation, and should be abandoned. But alongside all these people who mixed together the games of science and religion, and who rejected traditional Christianity in their speculation, you also have throughout modern history people who both contributed to modern science and also held to orthodox religious views, including Christians such as Rene Descartes, Blaise Pascal, Gregor Mendel, Roger Bacon, Galileo Galilei, Johannes Kepler, Gottfried Leibniz, Michael Faraday, James Clerk Maxwell, Asa Gray, Heinrich Hertz, Louis Pasteur, Lord Kelvin (William Thomson), George Washington Carver, Freeman Dyson, and Francis Collins. All of these people understood that in playing the game of science they were doing something different from the game of religion, and the play of one game did not nullify the play of the other.

The Rules of the Game of Science

In the preceding discussion of the history of the evolution of the modern game of science, I discussed two of the five rules for modern science, because they are most easily explained by that historical background: (1) Science is only concerned with the natural, with what human beings can sense, observe, and somehow measure within the known universe. Science is not concerned with the supernatural. The supernatural is, by definition, outside the bounds of science. And (2) the game of science is distinct from both the game of religion and the game of philosophy. The game of science looks at the truths of the natural world, and the game of religion looks at the truths of God, while the game of philosophy looks at the truths of the human reasoning process. Now I shall discuss three more important rules for this discipline of science. One of these is (3) the rule about repeatability, and another is (4) the rule about falsification. (Number 5 will be discussed after numbers 3 and 4.) These rules bring into the discussion the importance of what is called the scientific method, and the important role of scientific experimentation.

The scientific method is a logical scheme for understanding what is performed in the game of science. Different writers may phrase this in slightly different ways, but basically many would see five parts to this scientific method. There is the formulation of a **question** about some aspect of the world of nature; one or more **hypotheses** are generated to speculate about how the natural process concerning this aspect might work; one or more **predictions** are made about what might happen in connection with this natural process and the hypotheses concerning it; some kind of **test** or experiment is performed that allows the observation of the natural process in action (this test is designed as carefully as possible to isolate and expose one critical part of the process to determine whether or not the prediction based on the hypothesis was correct or incorrect); and the results of the experiment are then subjected to careful **analysis** to determine the significance of the results with regard to the predictions and the hypotheses, and with regard to answering the original question. If the experiment confirms the prediction, the hypothesis is considered more likely to be an accurate description of the natural process, and we conclude that we know more about natural science and are better able to predict and control what happens in the realm of nature. If the experiment does not confirm the prediction, the hypothesis is considered inaccurate, and it is set aside while new ideas are conjectured. Note that the experiment does not prove the hypothesis is correct; it merely increases the probability that the hypothesis is a good description of reality. For example, you can perform an experiment to see whether dogs prefer playing with balls or with cubes, and find that the dogs prefer balls. But this may be because of the prior experience of the dogs playing catch with their masters, and not because of any innate preference for balls over cubes. In such a case the experiment did not take all the possible factors into account and did not demonstrate the prediction based on the hypothesis, even though it confirmed the prediction. It is not at all uncommon for an experiment to neither confirm nor disprove a prediction, but only to lead to new questions and further speculation about the underlying natural processes. But the scientific method provides a system by which people can focus on specific questions and details, and learn more about what is true in the game of natural science.

However, it has been noted that in the history of the development of natural science, this method seldom has been the actual process that was followed. In many cases scientific discoveries were the chance results of experiments that were not seeking to investigate those particular details. The experiments were designed to investigate other details, and the results of the experiments provided data that no one anticipated, thus leading to new questions and new research. Therefore the scientific method does not really describe the process of scientific discovery, but rather it describes the logic of understanding the scientific data, and the process involved in examining and testing new data after it has been discovered.

The role of testing or experimentation is crucial in the game of modern natural science. It is no more important than the processes of hypothesizing prior to experimentation or analysis after experimentation, but it is just as crucial as these reasoning processes. The importance of experimental testing brings us to the third rule, concerning the role of repeatability. The game of science seeks to determine truth within the natural world. How do the forces and processes of nature work? Regardless of our beliefs about philosophy or religion, we all realize that the world as we experience it has a certain regularity built into its functioning. Gravity causes things to fall down; sharp objects cut things more easily than dull objects; dark clouds and wind call for umbrellas rather than sun-block. The game of science is not about the supernatural, by definition. In the game of science, we do not know whether a god created the natural world; all we know is what we can observe about the way the natural world works. The research of science is all about determining what factors bring about that regularity of operation. That regularity leads to the scientific **rule of repeatability: what one scientist does in an experiment must be able to be repeated by another scientist in another experiment.** (This will be Rule Three in the summary at the end of this chapter.) If I perform an experiment and I publish my results, then when someone else repeats my experiment and he gets different results, this suggests that either my experiment was somehow in error, or his repetition was in error (perhaps he was more sloppy than I was). Scientists participate in international conversation in order to help determine the truth of natural processes. Properly isolated, the same forces in the same experiment should produce the same results. The role of experimentation distinguishes the game of science from the game of philosophy. It is not enough to logically think through what the natural process might be. Your thinking must be matched by some experimental process that yields data that can be analyzed. The game of science is about observing the natural world, which means you must carefully define and measure your observations about what happens in the natural world. And that happening or event must be regular: it must be repeatable, either in the laboratory or in the observations of other people. If it does not describe the regular behavior of natural phenomena, then it is not science. There are no miracles in science, if by the definition of "miracle" we mean something happening outside the functioning of natural processes. (The term "miracle" can also be used to label a wonderful event, and in that sense science is full of miracles.) The concept "the regular behavior of natural phenomena" includes all the interactions of all the aspects of nature, including all special cases that may be statistically rare occurrences. For example, gravity always causes water to run downhill. When water is trapped inside a pipe system such that a siphon effect is created, so that the water runs uphill, this is a result of the natural phenomena of different pressures at the ends of the pipe system, and it is still governed by the principle that gravity causes the water to run—ultimately—downhill.

The role of experimentation also highlights the fourth rule, that of **falsifiability**. This label is perhaps the most common label used for this rule, but some scholars prefer to call this the **rule of testability. If a hypothesis leads only to predictions that cannot be tested, it is not considered scientific.** Rephrased, this says that **if a hypothesis leads only to predictions that cannot be in some way observed and measured in any experiment, so that there is no way of proving the hypothesis false, it is not scientific.** (This will be Rule Four in the summary at the end of this chapter.) I stated above that if an experiment confirms a prediction, this makes the hypothesis more likely to be a good description of natural truth. But it does not prove the hypothesis, for there may be more factors at work than the experiment was designed to measure. On the other hand, if the experiment was properly designed and carried out and it does not confirm the prediction, that effectively disproves the hypothesis (or at least that version of the hypothesis). Because there is always the possibility of other factors involved, perhaps even forces as yet undiscovered by human research, it is very difficult to prove a scientific hypothesis to be absolutely true. But it is easy to prove a hypothesis is not true if the results of experiments do not fit the predictions of the hypothesis. Therefore a major criterion in the functioning of the game of modern science is whether or not a scientific hypothesis can be proven wrong—whether it can be falsified. In other words, can you design an experiment that tests whether it leads to confirmable predictions? If you cannot design any experiment to test the idea, if there is no way to prove the hypothesis false by any experiment, then the idea remains in the realm of speculation; it is not science.

However, though these rules of repeatability and falsifiability are very important in the game of modern science, there are some practical limits to the application of these rules. The rule of repeatability has certain limits affecting it. For example, at the present time there is only one large hadron collider (LHC). It has been built and is operated by a collected group of the world's scientists with funding from several governments. If someone performs an experiment in this facility and produces results, those results cannot be duplicated in any other location because there is no other comparable scientific equipment. In theory a second similar collider could be constructed, but the financial cost means this likely will never happen. The best we can do is to have other people repeat the experiment in the same facility, and have careful analysis determine whether the results are genuine natural processes, or simply an accidental aspect of the facility. Some experiments are simply too complex or too expensive to replicate.

The limitation about replication applies in a different way to the science of archaeology: you can never repeat the process of digging up ancient remains. Once the first excavator uncovers the items, cleans them, and puts them in a museum, nobody can ever repeat his work to find the same results. In archaeology this has led to a methodology of intensive collection of field notes, photographs, and measurements during the excavation process, so the data can be re-analyzed at a later time. Also in the field of archaeology we find the popularity of the use of the stepped trench as a type of excavation. Instead of digging up an entire city, the excavators dig a trench that slices into the city mound in only one section of the mound, like a set of steps ascending the mound. That trench exposes parts of different walls and buildings, and uncovers valuable pottery remains. You can never redo that particular trench. But decades later you can shift to another part of the mound and dig a new trench, and you would expect to find similar walls and similar pottery types, so as to confirm the results of the previous dig. If you did not find similar items in a similar sequence, this would cause archaeologists to question any interpretation derived from the results of the first excavation.

This same limitation regarding repeatability applies to some degree to any exploration that is the first of its kind. We can only theorize about what some aspects of Native American cultures were like prior to the arrival of the Europeans in the Western hemisphere. Once Spanish, British, and French explorers began to move into the interiors of the American continents and explore the cultures they found, the results of the contamination from the earliest European contacts were already at work. Diseases previously unknown in the Americas had spread, and the cultural and political economies had been disrupted, as results of the presence of Europeans installed in the coastal areas. We can never go back and observe all the facets of the Native American population prior to the arrival of European explorers. We have tried to take this limiting factor of first-time events into account in the recent explorations of the space program. We have expended great effort to attempt to ensure that the rovers on Mars were purified of all traces of Earth-based microbial life prior to sending them to the other planet. If we did not do this, then should any future spacecraft find some trace of microbial life, we would never be able to know if such microbes were original to Mars, or were the result of contamination from Earth.

In a similar way, there are certain practical limits to the rule of falsifiability or testability. In theory certain experiments are possible, but in practice certain procedures are off-limits, or out-of-bounds, in the game. For example, we place limits on what kind of research we are willing to do on human subjects. We consider it immoral and improper to subject humans to certain kinds of medical or psychological experimentation. We can perform similar experiments on animals and we can reason from those results what the effect would be on humans, but even in that case we are experiencing conflicting opinions today about the propriety of such experimentation upon animals. After the invention of nuclear weapons, several governments signed treaties in which they banned further detonation of such weapons in the atmosphere. We consider the limitation in learning more about the effect of above-surface detonations to be small compared with the benefit of keeping radioactive materials out of our atmosphere. And as with repeatability, so too with testability: some experiments may be possible in theory but are simply too expensive or complex to carry out in practice. Drilling a hole to the center of the Earth constitutes the plot of many a science fiction story, but the actual drilling of such a hole is beyond our current ability.

The last rule for the game of science is that you have to be "in the game." Anything that you do outside the game does not count. In the game of American football, the field goal kicker might warm up by going off into another area where he practices kicking several balls toward a goalpost. He might have some very good kicks, perhaps even one successful kick over 60 yards. But that does not score any points. It does not count for anything, because at that time the kicker is not in the game. He is just off on his own, warming up. Only the actions that he performs while he is in the game count.

In the same way in the field of modern science you can explore and perform experiments and learn many things, but those actions don't count unless you are "in the game." To be in the game in modern science means that you have to be involved in conversation with other scientists. **You have to publish your results in some manner that makes the information available to other scientists.** (This will be Rule Five in the summary at the end of this chapter.) Then they are able to discuss this information, critique it, and do other experiments that test your results. If you just do your own experiments or explorations, and you do not somehow publish your information, you are not part of the conversation, not "in the game," and your results do not become part of the accumulated knowledge base of modern science.

In the history of modern science, we often identify one person as the first inventor or discoverer of some item, but we also note that other people performed similar work and made similar discoveries. The difference is that the one person credited as the inventor or discoverer either published, or perhaps obtained a patent, in a manner that made the information easily available to others. As an example, take Christopher Columbus. He is credited with being the first European to discover the continent of North America. Today we also know that long before Columbus the Vikings had crossed the north Atlantic and visited North America, and even founded colonies. However, those colonies did not survive, and the Vikings did not share the information about their discoveries with the rest of Europe. So that was private information among the Vikings, and soon forgotten even among the Vikings. It did not have any effect upon the knowledge of American geography in the rest of Europe. The history of the knowledge of American geography did not advance until Columbus and other explorers traveled beyond Europe and then reported the results of their travels to the rest of Europe in the late 15th and 16th centuries.

The game that we know as the modern field of natural science began in Europe about five hundred years ago, but is now shared in every nation. Scholars all around the world today participate in the research that is part of modern science, and they publish their results in international journals that are shared across the globe. There are attempts to keep some discoveries secret, either by private business firms, or by governments. But these discoveries cannot be used unless the information is made available to those within the business firms or governments, and that interior publication almost always becomes leaked to the rest of the world. The invention of nuclear power and nuclear weapons is a good example of this. The government of the United States financed a program to develop nuclear power and nuclear weapons during World War II. The American project succeeded in developing nuclear bombs by 1945. The Soviet Union was able to make its own nuclear weapons by 1949, even though the United States tried to keep this technology secret from the Soviet Union. The United States launched its first nuclear-propulsion submarine in 1954; the Soviet Union was able to launch their first one in 1958. Today nuclear power plants are used around the world, and there are ten nations that are believed to have developed nuclear weapons (one, South Africa, later disassembled its bombs and abandoned this type of armament). Even though the details of nuclear weapons are closely guarded secrets, it is understood that any nation with enough funding could develop them today. Political pressure and the lack of any practical purpose for such weapons are used to prevent other nations from constructing these bombs. This history illustrates that once a scientific discovery is published within a scientific community, that science is essentially available for all to use.

Here is a summary of the rules of the game of modern science.

1) Science is concerned only with the natural, not with the supernatural. By definition of what this modern field is, it cannot include the supernatural, or comment on it. Science is concerned with the regular and predictable processes of the natural world, and seeks to determine and understand those regular processes. When science investigates an exceptional event, something which some might call a miracle, it only seeks to determine what natural processes combined to produce that exceptional event. Science can never confirm that something supernatural occurred, nor

would it have any interest in that possibility. Science may investigate what some call the paranormal, concepts such as telepathy or telekinesis, but only in order to determine what might be previously unknown natural forces. Science cannot determine that such concepts are supernatural, or label them as such.

2) Science uses reasoning, but science is not the same as philosophy. Just as the study of pure mathematics is not the same as the study of engineering, so the study of philosophy is not the same as the use of reason in science. The process of experimentation and the rule of testability keep the game of science grounded in the investigation of natural phenomena. Imagination is extremely helpful in reasoning in the game of science, but the rules of science restrict the playing field to observable and testable aspects of the natural world.

3) Science requires the process of experimentation and insists on the rule of repeatability. The same processes and same situation must always produce the same result.

4) Science insists on the rule of testability, or falsifiability. While there may be some practical limits on what kinds of experiments we can perform and thus what we can test, science moves forward by continuing to design new experiments that enable the testing of more predictions derived from the hypotheses or theories that scientists have produced. [Theories are large collections of logical hypotheses that fit together in some consistent scheme.] If a hypothesis cannot be tested in any way, to determine whether the hypothesis is "true" or "false" in its assertion of how some aspect of nature works, it does not fit within the game of science.

5) Science requires the publication or sharing of information. You must contribute your data and your reasoning to the larger pool of scientific knowledge, and you must allow your results to be analyzed, tested, and critiqued by other scholars.

Chapter Four: The Rules of the Game of Religion

Corresponding to the five rules that guide the investigation of natural science, we can define six rules that guide the religious life. These rules are:

- 1) faith,
- 2) listening,
- 3) faithfulness,
- 4) trust,
- 5) fellowship, and
- 6) prayer.

The first rule is that the person must have faith that the supernatural exists, or that God is real ("And without faith it is impossible to please God, because anyone who comes to him must believe that he exists and that he rewards those who earnestly seek him" Heb. 11:6). If you do not think that there is any god, or anything supernatural, you are not going to detect it, and you are not going to heed it or cooperate with it. This rule includes having a sense that there is more to life than just atoms and existence. There is some kind of meaning, there is some kind of value. This includes a sense that what we do as humans matters. It matters somehow, to someone, somewhere. There is such a thing as value and meaning. There is some type of reward for good and bad behavior. But of course there is no judgment without someone serving as judge. Who is this god or spirit that is ultimately going to judge your life and the decisions that you made?

The second rule is that you must listen to God. ("Let the person who has ears listen!" Matt. 11:15[4]) If there is a god, and if he is trying to communicate, via actions or words or miracles or other interventions, you must tune in to what he is doing and pay attention. More will be said about this listening to God in the next chapter.

Third, you must practice faithfulness, that is, obedience. You can label this as discipleship. But the point is that you must actually respond to the information from the supernatural world. Or to put it another way, when Jesus says to you, "Follow me," you must actually follow. It is one thing to believe that God exists, but it is another thing to actually respond in obedience to God's instructions. Where there is no discipleship, there is not much of a religious life. (Cf. the epistle of James in the Bible: faith without works is dead—James 2:17.)

Along with faithfulness, you must have trust in God. This fourth rule is something more than faith in God's existence, or faithfulness to his instructions. This involves faith in his intentions and in his care. When you sense that God is instructing you to do something that is in some way dangerous to your safety or status, you have to trust that God knows what he is doing, and that he is going to take care of you. You have to believe that he knows what is best, and that he is guiding you in the proper way for the fulfillment of your life, whether it seems logical or not. (In Psalm 56:3 the poet states: "Even when I am afraid, I still trust you." In Acts 27:21-26 St. Paul predicts to his fellow sailors that they will experience shipwreck, but that no lives will be lost, and he offers

---

[4] *God's Word Translation.*

them words of hope in verse 25: "So have courage, men! I trust God that everything will turn out as he told me.")[5]

The fifth rule is the matter of fellowship. ("Everyone who loves the Father loves his child as well" 1 John 5:1) If you believe that God exists and that he is doing something in your life and in this world, then you have to care about what he is doing, and care about what he cares about. And that means caring about the people with whom he is involved. You must locate these people, and you must care for them. You must seek their company, and surrender yourself to their care for you. Even the Lone Ranger had his faithful companion Tonto. And in the TV series the Lone Ranger spent his time and energy seeking where other people needed help, the kind of help he could provide, and then providing that help. If you are going to walk with God, you must walk with his other disciples.

And then sixth, there is the matter of prayer. You have to talk to God. This could be restated for some types of religion as the fact that you have to consult the supernatural, by means of whatever method the supernatural world has opened a channel, be it astrology or tarot cards or crystal balls. But for the two biggest religions, Christianity and Islam, this simply means prayer to God. There are two types of prayer: public and private, and both are necessary. Public prayer is often more identified by the label worship. But the essence of public prayer is that you talk to God in the view of other persons. You are not ashamed to be identified with God, or with his disciples, and you demonstrate to the world how you rely upon God and how you believe that you, and others, can communicate with him. But public prayer must also be supplemented by private prayer. Prayer is not just

---

5   Both passages from *God's Word Translation*.

something you do in public with other people. You must develop your relationship with God by means of a regular private communication, sharing your concerns, expressing your private emotions, and thanking him and acknowledging his presence and help. ("To you, O LORD, I lift my soul. I trust you, O my God. Do not let me be put to shame. ... No one who waits for you will ever be put to shame ... Make your ways known to me, O LORD, and teach me your paths. Lead me in your truth and teach me, because you are God, my savior." Psalm 25:1-5)[6]

If modern natural science is concerned with interpreting and interacting with the natural world, religion is concerned with interpreting and interacting with the supernatural world. These six rules lay out the basis of interacting with the supernatural world. The next two chapters will focus on the proper method of interpreting communication from the supernatural world, on the matter of how to know what God is saying to you.

---

[6] *God's Word Translation.*

Chapter Five: The Importance of Sacred Scripture in Modern Religion

A Brief Discussion of Types of Religion

Scholars of religion today understand that there is no one definition of the concept of religion that fits all the types of religious phenomena that are found in human societies. Some belief systems include many of the aspects linked to religion, but dislike the label "religion." Marxism is one example of this. Marxism is a system of beliefs about the role of humans in the world, about the morality of human interactions, and about the structure and purpose of human society. It includes a belief about the past and about the future destiny of human society. But Marxism denies the existence of any god, or even the existence of the supernatural. Therefore it rejects the label of religion, even though it is best understood academically as an atheistic religion. Buddhism is another religion that, though it does not completely reject the role of gods, has little use for them. Buddhism is a belief system that focuses on the individual's integration of his understanding of existence with the ultimate truth of the universe. Buddhism includes a system of morals that are believed to help affect the progress of the soul of the individual. The central teachings of Buddhism aim to help the human soul achieve a happy eternity beyond the point of death. Buddhism certainly believes in the world of the supernatural. One way to understand Buddhism is that it teaches that this life is a temporary illusion and that a person should learn this truth and shake off this illusion, thus moving to a whole different existence. The term "Buddha" actually means "the one who has awoken to the truth." These two examples, Marxism and Buddhism, illustrate that religious systems can vary considerably. You can have a religion that denies all existence of the supernatural, or at the other extreme you can have a religion that denies the reality of the natural world, teaching that the natural world is only an illusion, a dream from which one's mind must awake.

In this book I am concerned about the relationship between science and religion, which I can rephrase for the moment as the relationship of the natural and the supernatural. From that framework, I can categorize different types of religion according to how they view the relationship of the natural and the supernatural.

Ancient and Tribal Religions

Some religions identify the processes of nature with the supernatural. Many aspects of ancient tribal religions are essentially imaginative interpretations of the natural world, with the forces of nature conceived of as spirits, and the forces of tradition or human society conceived of as other spiritual powers, perhaps ancestral spirits. Many of the great religions of the ancient world viewed things this way, such as the religion of the ancient Greeks, or Egyptians, or Babylonians. The sun was thought to be a manifestation of the god who brought light and heat, the moon was another god, there was a god who brought rain and lightning, and another god who ruled the deep sea. These great religions of the ancient world no longer have any devotees because our expanding understanding of natural science has reclassified these phenomena as natural items that function according to natural laws which we have learned to understand. Since these phenomena are no longer part of the supernatural, the phenomena ceased to be understood as gods with personalities, and the religious systems ceased to have meaning and died away.

A similar extinction happens with modern tribal religions today when contact with the developed world intrudes upon their lives. Modern technology brings significant changes to their world. Modern medicine brings better healing from disease; the power of modern weapons, transportation, and communication changes the understanding of the control of the world and of their society; the economic situation is completely changed and often the physical territory is changed as the developed world moves in to take control of natural resources. Under the force of such changes those tribal religions often collapse, or undergo great shifts such that they become merely a psychological refuge for some tribal members. Such members cling to the old traditional beliefs and deny the significance of the benefits of modern science.

Religions that Focus on Mental Powers

Some religions focus entirely on the human mind or consciousness. The natural world is considered to be of much less importance, ultimately of no importance. Some modern religions such as Scientology or modern adaptations of Hinduism or Buddhism go in this direction. They teach that the only things important are those things that happen within your own mind, and how you choose to perceive things. The functioning of the mind is where one moves beyond the world of the natural, into the realm that is above nature. The mind, properly trained or enlightened, becomes capable of supernatural activities. In some cases the soul or mind is believed to be capable of separating from the body, to travel to other locations and to hear and see distant events. In some cases these other locations are in other worlds, or in places such as heaven or hell. Some of these religions offer the hope that a person can learn to use the power of the mind to influence or change things in the natural world. The power of the mind may perhaps be able to heal diseases, or to move physical objects without any natural contact. Or perhaps, like Darth Vader in *Star Wars*, one can use purely mental energy to choke the life out of some foe.

Social Religions

Other religions can be categorized as social religions. These are belief systems that are wrapped around the importance of the community and maintaining the tradition of the community. They deify the status quo, and view as a threat anything that would seek to change the social community. Such systems may cite the motto "God and Country," but in effect the "country" defines what "god" is. These religions will talk about supernatural power and supernatural beings, but such words boil down to being code words for defining the current social system. In these religious schemes the supernatural is located in the ideas that govern society. Alternatively in the view of some radical movements, these code words may define, not the current society, but some desired future social system. Thus in history we find some people using force to defend the capitalist economic system with the separation of owners and laborers, and others using force to work for the Marxist communist economic system. Both groups feel they are involved in something holy and may speak of doing the work of God, such as in the words of modern Liberation theology. Liberation theology is a version of Christian theology that has adopted much of Marxist doctrine into its system in order to justify opposing the established economic system. A similar view provides significant insight on the problem presented by the rise of radical Muslim groups in the Middle East. You have the traditional Muslim dictatorships working to defend their power and position, in the name of God, who are challenged by the radical Muslim groups seeking, in the name of God, to bring into existence their ideal view of a different type of Muslim society; and all the while the vast majority of the world's Muslims who live outside the Middle East pay little attention to that regional struggle and go about their lives in their own different countries.

Religions that Recognize both the Natural and the Supernatural

The two largest religions in the world today, Christianity (33% of the world's population) and Islam (22% of the world), are religions that believe in both the reality and importance of the natural world, and the reality and importance of the supernatural world. They view the natural and the supernatural as distinct and coexisting, with each realm having some effect upon the other, because God so chooses. God created the natural world and designed it to run according to natural laws that he established and upholds, and God rules the supernatural world wherein there are other spiritual beings that he created. God is active in the natural world, both by upholding the natural laws he established, and by special intervention in those processes from time to time, as he chooses. The actions of humans in the natural world have some impact upon the supernatural world for two reasons: first, God chooses to care about the humans and the world he created, and thus the human actions make God, and his angels, either happy or sad; and second, God has chosen to operate such that these human actions become a factor in determining what God does in the future, including the relationship of the natural and supernatural realms. In other words, a human action such as repentance from sin influences what God does in both the natural and supernatural worlds, both now and later.

In this book I want to focus on the discussion of the relationship between science and religion with regard to those types of religion that hold to this coexistence of both the natural and supernatural worlds. In the chapter on natural science I explained that the separation of the study of the natural from the study of the supernatural began long ago when humans began to analyze the cause-and-effect functioning of various regular processes of nature. This study of natural processes was greatly advanced by Muslim scholarship during Europe's Medieval period.

The distinction of these two areas of study was clarified intellectually by the reasoning begun with Descartes and others. This type of scholarship of the natural world was further advanced in Europe during the period called the Enlightenment and continues now in the modern world. If science is defined as the study of the natural world, then by this definition the study of the supernatural world is something else, the domain of religion. Science rightfully investigates and makes determinations on everything observed in the natural world. But science cannot observe or determine anything about the supernatural. The supernatural is, by definition, that part of truth or reality that is beyond science. It is beyond science because it is beyond human observation or human investigation. Because it is beyond human observation or investigation, atheistic scientists or philosophers would be correct in rejecting the supernatural as a source of truth, or as a source of any information, *unless* the supernatural world in some way reached into the natural world and made itself known. In the intellectual framework I am describing here, humans can know nothing of God, unless God himself chooses to make something known. But this is the heart of the large modern religions of Islam and Christianity: God has spoken, God has intervened and revealed additional information, additional truth. This matter of revelation from the supernatural world is also a significant part of many other religions, such as whenever a shaman or medicine man contacts the supernatural world and brings back a message. God, or The Great Spirit, or however the religion conceives of supernatural power, has spoken, and continues to speak. God intervenes by acting within the natural world. Therefore it is the role of the game of religion to deal with this acting of the supernatural, or this speaking by God. Religion deals with supernatural revelation.

This conception sets up a clear divide between these two games. **Science cannot speak about religion, and religion cannot** speak about science. **It is not the role of religion, as limited and defined for this discussion, to investigate or discuss science. The role of religion is to grapple with the supernatural.** And just as the game of science has its rules that we must know and follow, so too the game of religion has its rules that we should know and follow.

If we go back to the illustration of playing checkers on a checkerboard as an example of the game of science, I pointed out that the red squares are not part of the game. But for religions that focus on the supernatural, the red squares do exist, alongside the black squares of the natural world, and we can think of some of the red squares as containing written instructions for how to play the strategy of the game. They tell you where to move on the black squares. Revelation consists of the information on the red squares being shared with the players on the black squares.

A Brief Review of the Interpretation of Supernatural Revelation

How does the supernatural world intervene into our world of perception? We can define three methods: religion can intervene by some type of *action*, or by some sort of *symbol*, or by a direct message of human *words*. But in order to be understood, this supernatural intervention must be interpreted by the human viewers. The act of interpretation requires a human interpreter.

But I assert that the action of human interpretation of the divine intervention is significantly different when the intervention consists of human words, rather than some type of action or symbol. The interpretation of some divine *action* or some *symbol* depends upon having a special human who is qualified to perform that function. The interpretation of human *words* does not require a special human, but is something any human can do by applying the ordinary rules of interpreting language.

Interpreting Supernatural Actions

Interpreting the divine or supernatural action intervening in human history has always been the domain of religion. Let me give three examples of rain, war, and disease. The first example is rain: when the regular process of rain stops and we experience an unusual drought, humans have almost always asked the question of why God has done this. The same is true of the other extreme of floods from too much rain, or when we experience unusual storms such as hurricanes or tornados; these mysterious forces were believed to be controlled by the gods. Even today in the legal language of insurance contracts such unpredictable events as floods and natural fires are often referred to as "acts of God." The second example is war: although generals do their best to guarantee the outcome of each war, wars do not always go as planned. "For want of a nail, the shoe was lost …" goes the old rhyme, pointing out that the most minor detail can in some cases cause a complete reversal of expectations. [7] This is often perceived as an act of God: he makes the most minor of changes in a tiny detail, and thus controls the outcome of great human endeavors such as wars, sometimes in defiance of human expectations. And the third example is disease: one person catches Asian bird flu and dies; 80% of the people infected with Asian bird flu experience no symptoms and never know they were infected. One person smokes tobacco all his life and lives a

---

7   Traditional proverbial rhyme with obscure source:
    For want of a nail the shoe was lost. / For want of a shoe the horse was lost.
    For want of a horse the rider was lost. / For want of a rider the message was lost.
    For want of a message the battle was lost. / For want of a battle the kingdom was lost.
    And all for the want of a horseshoe nail.

vigorous life to age 90; another smokes and dies of cancer at age 40. These must be the decisions of God.

But there is a problem with understanding these actions as supernatural interventions in history: who gets to decide what it means? Furthermore, who gets to decide when the unusual action is indeed a special intervention from the supernatural world, and when it is just a matter of probability in the functioning of the natural world? In the Christian Bible we read that God "sends rain on the righteous and the unrighteous" (Matt. 5:45). Obviously this means that God also sends lack of rain, or drought, upon both the righteous disciple and the unrighteous non-disciple as he chooses. So unless someone has some other additional revelation, how can one interpret the matter of rain as a special act of God? The same problem applies to war: was victory or defeat a matter of supernatural intervention, or just chance, or just good or bad human strategy? Who gets to decide what the result means? The same problem applies to the interpretation of disease. Many religious scriptures affirm that their gods often allow evil people to prosper and live long happy lives (see the book of Ecclesiastes in the Bible, for example), while on the other hand the gods may also send suffering to good people for reasons not revealed to the people (in the Bible note the story of Job). Unless you have some special person with some additional insight, you cannot interpret the meaning of these actions.

Many religions have just such a special person. These people are labeled prophets, or shamans, or wizards, or soothsayers, or oracles, or diviners, or witch doctors, or mediums, or priests, or other similar titles.

With regard to the Jewish and Christian traditions, we can point to the important role of God's prophets and priests as these authorized interpreters. For example, when a prophet such as

Elijah or Amos said that a drought was a punishment sent by God, then that interpretation was official (1 Kings 17-18; Amos 1:2, 4:7-8). But these interpreters provided their instruction about the meaning of the special event on the basis of *other, additional revelation* that they had received. They somehow had access to know the mind of God. They had the ability to tell you what God was thinking.

Interpreting Supernatural Signs

The role of the authorized interpreter of God's action in history also functions in the second case of revelation by symbol. In many cases the will of a god is believed to be revealed by some sort of signs or symbols. An African medicine man will cast lots and produce an interpretation for the pattern of the fall of the lots. A Babylonian priest would inspect the liver of a bird to find the will of his god revealed in the pattern on the liver. An Eskimo shaman will read the pattern of the cracks in an animal bone to determine the information from the spirit world. A Chinese medium will interpret the symbols traced on a table by a person who is channeling information from the supernatural world. The ancient Israelite priest would utilize the Urim and Thummim to determine whether God replied "yes" or "no" to some question (Exodus 28:30, Numbers 27:21, Ezra 2:63). A modern mystic will turn over the tarot cards and interpret the meaning of the card sequence. A palm reader will find your entire future planned by your god and printed in the lines on the skin of your hand. In all these cases we have someone who is believed to have some sort of authority for reading and interpreting the symbols from the supernatural world. This is parallel to the interpretation of divine actions. These are all illustrations of the role of the person who is considered to have the authority to interpret communication from the supernatural world, either by interpreting the symbols, or by interpreting the historical events.

These authorized interpreters of the supernatural world often have to train for years with an older interpreter, and often have to have some confirming sign from the supernatural world that they have been so authorized. Thus one person must train for years with a medicine man before he can work alone, or a person must train with other priests before he can serve alone. With regard to a confirming sign, sometimes a person will have a series of dreams in which specific things happen, and the person and his culture recognize those dreams as being the proper sign. Sometimes the person will hear a voice, or see some other communication from the supernatural world, and recognize that his god is calling to him. His account of this event will be recognized by others as displaying the proper authority. Sometimes a person will be linked in some way with what is perceived as miraculous activity: someone will be mysteriously healed from severe illness through this special person's ministrations, or during a drought water suddenly will be found as a result of his efforts. In other cases the confirming sign is merely the official authorization by previous religious leaders. A ritual such as ordination is considered to establish the validity of the ministry of the new interpreter.

In all of these cases, a certain practical test works itself out in the history of the service of the interpreter of the supernatural: what this person says comes true. If this person makes predictions about the future, those predictions are fulfilled. If this person offers supernatural help to provide healing from illness, rescue from drought, or safety from some specific threat, this help is achieved. If this person delivers a warning or threat from the supernatural world, this threat comes to pass in history. If the words of the supernatural interpreter do not come true, then his authority is rejected. In that situation the community generally discounts that person and rejects his claim to knowing the supernatural. Note this example from the Bible, Deuteronomy 18:17-22:

> The LORD said to [Moses]: "What they [the Israelites] say is good. I will raise up for them a prophet like you from among their brothers; I will put my words in his mouth, and he will tell them everything I command him. If anyone does not listen to my words that the prophet speaks in my name, I myself will call him to account. But a prophet who presumes to speak in my name anything I have not commanded him to say, or a prophet who speaks in the name of other gods, must be put to death. You may say to yourselves, 'How can we know when a message has not been spoken by the LORD?' If what a prophet proclaims in the name of the LORD does not take place or come true, that is a message the LORD has not spoken. That prophet has spoken presumptuously. Do not be afraid of him."

In the case of Judaism and Christianity, and continuing into Islam, this principle held: the prophet was recognized as truly sent by God only if his prophecies came true. This could only be judged by short-term prophecies, of course, for the long-term predictions would require decades or perhaps even centuries to come to pass. But God promised to validate his prophet by short-term prophecies so that the people could be confident in God's communication through this particular prophet. This is part of the reason why the Biblical God became so angry during the Old Testament era when the people of Israel rejected his words, because he had gone to the trouble of carefully validating his prophets.

But we know that there are different religions in the world. A person recognized as an authority regarding the supernatural in one religion is not necessarily recognized as an authority in another religion. The ancient Israelites were told to reject the words of the Canaanite prophets and preachers, even if their predictions came true: "If a prophet, or one who foretells by dreams, appears among you and announces to you a miraculous sign or wonder, and if the sign or wonder of which he has spoken takes place, and he says, 'Let us follow other gods' (gods you have not known) 'and let us worship them,' you must not listen to the words of that prophet or dreamer. The LORD your God is testing you to find out whether you love him with all your heart and with all your soul. It is the LORD your God you must follow, and him you must revere. Keep his commands and obey him; serve him and hold fast to him" (Deuteronomy 13:1-4). Muslims and Jews today reject the authority of the Christian apostles who wrote the New Testament, and Christians and Jews reject the authority of Muhammad who delivered the Quran. Christians, Jews, and Muslims are supposed to reject the authority of modern spiritualists and prophets of new religions. In actual practice, many believers in these three major religions continue to be attracted to the information provided by astrologers, palm readers, spiritual channelers, or other preachers of different revelation. Some Christian preachers today rely on their special experiences to validate their credentials as speakers for God, rather than on their ordination by the church. The issue of who has the authority to interpret either a god's communication through symbols or a god's interventions in history becomes a part of the faith. Different religions recognize different human authorities. In Christianity, the Roman Catholic Pope claims the ability to speak with divine authority as necessary to clarify the will of God in the modern world. Protestant Christians and Eastern Orthodox Christians reject this claim. The Mormon Church claims that its leaders have a similar authority to speak for

God as necessary.[8] Mormon doctrine is rejected by traditional Christians of all types, as well as by Jews and Muslims. Those people who follow modern spiritualists, shamans, or priests of other religions generally reject the authority that is claimed in the Jewish, Christian, or Muslim religious faiths.

So the matter of interpreting the communication from the supernatural world by way of prophet or priest, by way of some specially authorized human interpreter, who interprets either a divine action or a special symbol, is complicated by the difficulty of recognizing who has that special authority. The truth is that anybody can claim to have such authority. It is, unfortunately, not uncommon for someone to say that God, or the spirits, or someone from the other side in the supernatural world has communicated to him and therefore we should give heed to that person. In fact, while some people are deliberately lying and trying to fool others into following them, in many cases the person making such claims genuinely believes that the supernatural world has reached out to him in communication. So how are people today to know who to believe? I stated above that one test was whether or not the predictions of the prophet came to pass. But in Deuteronomy 13 that test was described as only partially valid. The other test was whether or not the supernatural spokesman was pointing the people to follow the already accepted God, or was pointing in a different direction. In other words, the new revelation had to be consistent with the previous revelation, and not contrary to it. But this test requires that there be some previous revelation. For the ancient Israelites, all subsequent prophets had to be consistent with the teachings

---

8   The Church of Jesus Christ of Latter-Day Saints has recently chosen to cease using the label "Mormons/Mormon Church." Given the long history of the use of this name, it is not clear that this desire to change the common name will be effective.

of the great prophet Moses. The words of Moses were enshrined in the Torah scrolls. Thus the new words of revelation had to be tested against the written words of the past. And with that we had the beginning of the tradition of the written scriptures of revelation. I stated above that the interpretation of divine Scripture is different from the interpretation of divine action or some divine symbol. Now I need to explain this further.

The Role of Scripture as Divine Revelation

Many modern religions have a collection of written scriptures that are considered to have divine authority. This is often called a "bible" today from the ancient Greek word for "book," and it is often dignified with a capital *B* when referring to a specific collection trusted by a specific religion, such as the Christian Bible. The supernatural world has somehow reached into this world and created this deposit of documents for the benefit of the religious believers. Judaism has the Tanak making up its Bible (the equivalent of the Christian Old Testament), Christianity has the Old and New Testaments making up its Bible, and Islam has the Quran. The Sikh religion has its scriptures, and the Baha'i religion has its holy books. The Mormon Church has a "third covenant" (or "third testament") in which it adds three books to the Christian Bible: these are the *Book of Mormon*, the *Doctrine and Covenants*, and the *Pearl of Great Price*. Modern Hinduism has a large collection of various writings dating to different eras, which are considered by different groups within the religion to have divine authority. Some examples which find wide acceptance among Hindus are the Vedas, especially the *Rig Veda* (hymns to the gods, something like the Psalms in the Christian Bible), the *Ramayana* (the story of the great Lord Rama), and the *Bhagavad Gita* (major teachings of the great Lord Krishna). Buddhists accept the *Tripitaka* as sacred scripture, with different Buddhist subsets also accepting other texts. It is fairly clear that modern religions which have developed in the historical period following the rise of Christianity have been influenced by the Christian concept of sacred scripture. It is difficult to assess the degree to which more ancient religions shared this concept, due to the scarcity of appropriate historical resources. In the case of modern expressions of these ancient religions, such as modern Hinduism, it is likely that the concept of the Jewish, Christian, and Muslim scriptures has influenced how these other religions view their scriptures today, and how they understand the history of the view of their scriptures. It is notable that one of the efforts of many tribal religions as they seek to preserve their faith and culture today is to compile a set of written records that will contain their authoritative tradition.

The existence of a collection of sacred writings for a particular religion is something different from the matter of revelation from the supernatural world through other sacred symbols. A communication by symbols has to be explained by some official interpreter in order for people to understand the divine message. A collection of written human words must also be interpreted, but interpreting written human language is different from interpreting signs and symbols. Symbols must be placed into some kind of meaningful context and then be assigned an interpretation. As one example of the interpretation of symbols, consider the story in the Christian book of Revelation where John sees in a vision a lamb which appears and goes to sit with God on the throne in the center of heaven (Rev. 5). This set of symbols is easy to interpret within the context of Christianity. The "lamb of God" is a frequent identification of Jesus; Jesus is the divine Son of God who came down to Earth from heaven, and after his ascension resumed his place with his Father on the throne of heaven. Thus Jesus is now King and Lord of the Universe, and Jesus controls all that happens, according to Christian theology.

For another example, consider when a Native American watches an eagle fly across the sky, land in a tree, look around, and then fly off in a certain direction. The Native American may interpret that event within the context of his tradition, taking into account both the significance of the eagle and his own current situation. Perhaps he finds some indication of direction. If the eagle heads west, it may be interpreted as the Great Spirit telling the Native American to head west. Or it may be that the bird's temporary setting in the tree means that he is not to head west, but to wait for a time, and only later in his life should he head for the setting sun. With symbols, we need a context and an authorized human interpreter.

But the interpretation of a written text is different from that of symbols. The text does require interpretation. It may be written in a different language, which must be translated (such as from Greek to English), and it may contain words with symbolic significance (such as a lamb meaning Christ Jesus). But the final communication of the text is generally clear to any informed reader. We do not need a special authorized interpreter. In the case of the book of Revelation, Jesus, the Lamb of God, shares the throne of God in heaven. We do not need some special interpreter authorized by the supernatural world, such as a shaman or prophet or priest, to explain what the events or symbols mean. The written text is in human language, and it speaks in a logical sentence. The proposition expressed in the sentence is the meaning of the communication.

There are written scriptures that are more symbolic and that are more complicated to interpret and to understand. For example, in the Christian Bible the book of Canticles (also known as the Song of Solomon) is a poetic story of a love relationship between a man and a woman. This is generally considered to be an allegory, in which the surface story symbolically presents a greater theological story. One common interpretation of this book is that this love story symbolizes the love between God (the bridegroom) and the Church (the bride). This is not the only possible meaning for this part of the Christian Bible. Another interpretation offered by some scholars is that this book is advice offered to young women about the need to be very careful in the experience of sexuality. In this interpretation, the book of Canticles is placed alongside the book of Proverbs. As Proverbs gives advice to young men regarding sexuality, so Canticles gives advice to young women ("Do not arouse or awaken love until it so desires," that is, let the full enchantment of sexual love wait until its proper time, don't rush it; Cant. 2:4). But whichever way you choose to interpret this portion of sacred scripture, the fact that it is human language (even if it is in the form of poetry!) allows the reader to understand that Canticles teaches something positive about love and fidelity. The previous example of symbolism of an eagle sitting in a tree does not reveal any clear meaning to the ordinary observer who lacks a properly authorized interpreter. In contrast, the account of the Lamb of God taking his place upon the throne in heaven gains its full meaning from its context within the body of scriptures comprising the New Testament. If such a vision were observed without that context, you would not know what it meant. Perhaps farm animals are taking over heaven and the world is about to fall into chaos? But in the use of sacred scripture, even when the text contains symbols, the vision is described within a written context that allows for fairly easy interpretation by any reasonably educated reader.

Supernatural revelation by way of scripture is different from the other two types of intervention. For practical purposes, the first two cases—supernatural intervention in history, or supernatural sending of symbols—require an authorized interpreter. There must be some special human who has the ability to explain what the supernatural intervention means. But when you have sacred scripture, any educated person who can read the scripture can interpret what the human words mean. In scripture the god speaks plainly in human language.

One of the principles enunciated during the Protestant Reformation was that every believer has the right, and indeed also has the obligation, to interpret the scripture himself. The renewal of literacy in the Renaissance fostered a renewal of the study of the Bible in Europe. This study of the Bible led to the conclusion that some doctrines and practices had evolved in church history that were not in agreement with the scriptures. The Protestants thus called for the "reforming" of Christian doctrine and practice to become once again consistent with the Christian scriptures.

Unfortunately, because of historical circumstances involving finances and politics, this resulted in the reformers challenging the authority of high church officials who opposed such reformation. This led very swiftly to a direct clash between the claims of authority among special persons and the claim of authority within the special revelation of scripture. The Protestants set up scripture as their highest authority for church matters. This clash led to the division of the Christian church into more denominations. In addition to the older divisions such as the Eastern Orthodox churches and the Coptic Church, the western Roman Catholic Church gave birth to the Lutheran, Reformed, Anabaptist, and Anglican denominations.

This principle, that every believer had the right to interpret scripture himself, was often abused in later history and still is abused today. The principle did not mean that every person had the right to his or her own interpretation, and that there was no one correct interpretation that should be accepted by all people. The principle maintained that written scripture was conveyed in plain human speech, and that any educated reader had the right to read it himself and determine what it said. The text had its own meaning. An interpreter was not free to make up his own interpretation. He was required to determine what the text said. But no one could take away from him the right to read and interpret the text himself. No one, not even the highest church official, not even the highest academic scholar, had the right to say, "This is what the text means, and you cannot contradict me, you must accept what I say." If the believer found that the text said what the official stated, all was well. If the believer found that the text said something other than what the official maintained, the believer was obligated to reject the official's interpretation. The text had its own meaning and its own authority.

This is not the same situation as the interpretation of events in history or the interpretation of revelation by symbols. Those situations required an authorized interpreter. But revelation by scripture was different. Everyone who could read the scripture was competent to interpret what it meant, because it conveyed its message in ordinary human speech.

So the interpretation of special actions in history, or special signs in history, requires an authorized human interpreter. But in practice that interpreter must be shown to be consistent with the revelation of the past. That emphasis on consistency with prior revelation gives rise to the importance of sacred scriptures in most modern religions.

Thus the rules governing the game of religion include the rules governing the interpretation of religious scripture. This scripture contains information transmitted from the supernatural world into the natural world. The game of religion is the matter of the interpretation of the supernatural. Therefore the game of religion requires the interpretation of the sacred scriptures. I will discuss these rules of interpreting scripture in the following chapter.

For most modern religions with a set of sacred scriptures, it has become important to maintain a harmony between what those scriptures say and how the doctrines and practices are maintained and carried out in the religious community. Even in Roman Catholic circles the authority of the Bible has never been denied, and the official teachings of the Roman church are slowly edging closer to Biblical views. For example, the Second Vatican Council (informally known as Vatican II) in 1962-65 admitted that there are Christian believers outside the official records of the Roman church (i.e., Protestants are also Christians), allowed worship to be conducted in local languages, and promoted rather than repressed the use of the Bible. In 1999 the Roman church declared that Luther and the Lutherans were "not wrong" on the teaching of justification by faith before God, which was the central tenet of argument during the Reformation era. This need to maintain a harmony between what is read in the scriptures and what is taught in the religion extends then to the interpretation of historical events and the interpretation of symbolic revelation within the religion. When someone ventures to interpret an event in history as the supernatural intervention of a god, displaying the god's will in some way, that interpretation of the god in action has to be consistent with the way that particular god is revealed to think and act in that religion's scriptures. When someone attempts to interpret some sort of symbolism as revealing the will of a god, that person's interpretation will be examined for its consistency with the communication of the god as revealed in the scriptures. When someone risks such an interpretation of history or of symbols, and his interpretation is not considered consistent with the religious community's scriptures, that person's claim to have special authority to interpret such things is greatly diminished. Thus having and using a set of sacred scriptures creates a strong source of authority within a religion. It does not rule out the use of other authorities, but it tends to circumscribe the functioning of such other authorities. And conversely, when someone desires to attack a religion or to make a change within a religion, it is often the case that the primary task must be to undercut the authority of the written scriptures.

Let me summarize this chapter now, before I move on. Over half the population of the world today claims either the religion of Christianity or the religion of Islam. Both of these religions have sacred scriptures as their primary source of revelation from the supernatural world. Most other modern religions have sacred scriptures that play some similar significant role in the functioning of those religions. That is why in this book I have chosen to focus my discussion of the game of religion on those religions in which sacred scripture plays a fundamental role. For these religions, the convention has evolved in today's world that the interpretation of nature is the game of science, and the interpretation of the supernatural is the game of religion. In these religions the controlling focus of the interpretation of the supernatural is the interpretation of that religion's sacred scriptures. The interpretation of anything connected with the supernatural has to be consistent with the scriptures, and the interpretation is judged by those scriptures. In the previous chapter I stated that one of the rules of religion is that the person must listen or pay attention when the supernatural communicates something. Therefore the rules of the game of religion in the modern world require attention to the rules of the interpretation of the sacred scriptures.

Chapter Six: The Rules for Interpreting Written Revelation

The Rules for Religion with Written Revelation

So what are these rules for the interpretation of the sacred scriptures? I will divide my discussion of these into a set of ten rules.

Rule One: Literal Interpretation. One simple way of phrasing the point of this rule is that the text means what it says. The text is actual human language. The text is not written in some mysterious spiritual language that requires that God send a special interpreter. The words have their normal meanings, and the ordinary rules of grammar apply. Any person who can read that particular language can interpret what the text says and therefore what it means. If there is some special technical aspect to the subject matter, then it may require that the reader also have that type of technical knowledge. For example, if the text is providing information about the design of a temple, the reader may need to know the technical terms for door lintels and window sills, or for special types of beams and columns, etc. If the text is discussing warfare, the reader may need to know the technical terms for the different types of weapons or armor. But with that qualification, the text is intelligible. It needs to be interpreted in the sense that all human language needs to be interpreted. But it does not require any further "spiritual" interpretation by some religious official.

Rule Two: Context. Both the words and the sentences in scripture are found within a context of other words and paragraphs. If there is some uncertainty about what some words might mean, the answer is determined by the context. How are those words being used in the surrounding text of the scripture? How do the surrounding paragraphs describe and define the topic under discussion? The reader is not allowed to assign any meaning he chooses to the words of the text. The reader is required to follow the train of thought indicated by the writer within the totality of the scripture. In Christian tradition this rule is sometimes phrased as "Scripture interprets Scripture." This means that when you encounter a passage that is difficult to understand, and different possible interpretations are suggested by different readers, you use the larger context of the rest of the sacred scriptures as a control on the interpretation. The interpretation that agrees with the more clear meaning of other passages is the one to be used. An interpretation that disagrees with the more clear meaning of other passages is to be rejected. The reader must assume that the original writers and editors of the sacred scriptures were ordinary logical people, and that what they were saying and writing was intended to fit together in a logical framework. The reader is not free to take any part of the scripture out of its context and interpret it in some way that is at variance with the clear thought structure of the rest of the scripture. For example, when God speaks in the book of the prophet Isaiah to say: "I have no pleasure in the blood of bulls and lambs and goats ... your incense is detestable to me" (Isaiah 1:11-13), this cannot be interpreted to mean that this God hates all ceremonial animal sacrifice. That would contradict the passages in other books such as Exodus, Leviticus, Numbers, Ezekiel, and Psalms where the same God gives directions for ceremonial sacrifice and expresses appreciation for proper ritual. (And, if pressed far enough, such an interpretation would suggest that God hates and rejects the sacrifice offered by Jesus; that sacrifice is the central truth of the Christian faith.) This sentence in Isaiah does not mean that God hates all animal sacrifice, but that God specifically rejects the sacrifices being offered by those particular Judahites at that particular time in history, because, as the text indicates, their ritual obedience was not being matched by their moral obedience, and thus their ritual practice was in effect lying about their relationship to their God.

This rule about context also requires the reader to take proper account of the type of literary genre. If the particular text of the scripture being studied is part of a historical narrative, then it must be recognized as such and so interpreted. If instead it is part of a symbolic vision, the interpretation must take that into account. If it is part of a law code, that is, a list of laws for the population to follow, then the proper interpretation requires the reader to interpret it in the context of what we know about how law codes work in our culture and in others. Everyone knows that the technical legal language of contracts is somewhat different from ordinary speech; that's why you have to hire a lawyer to review the contract before you are ready to sign it. The same was true of ancient legal codes. When in the book of Genesis Abraham purchased the cave at Machpelah to bury his wife (Gen. 23), the recorded exchange of words is almost certainly a technical legal transaction of ancient society where the words and phrases held additional legal connotations that are not immediately obvious to the casual reader. The reader needs to bring in other information related to the legal possession of land in the ancient world. Another genre we sometimes find in the Christian Bible is copies of ancient letters. There we need to recognize the form of the letter, and take account of the details that are part of the form of the letter, and not necessarily part of the important message of the text. If the text being studied is in the form of poetry, then the proper understanding of poetry needs to be applied to the text.

Rule Three: Figures of Speech. A part of what is meant by the statement that we must use literal interpretation is that we need to take account of figures of speech. The literal meaning of any sentence is that meaning that was intended by the original writer and would have been understood by his original audience.

Human language is a very flexible tool, and we can use it to communicate many things in many ways. We all understand and use many different types of figures of speech in everyday communication. A literal interpretation is different from what is sometimes called a literalistic interpretation. In a literalistic interpretation figures of speech are ignored and the interpreter insists that words can only have their most basic meaning. Thus when Jesus calls Herod a fox in the New Testament, a literalistic interpretation would insist that this means Jesus is informing us that Herod is really a fox and not a human. But a literal interpretation recognizes a metaphor as part of ordinary human communication, and identifies it as such. As a metaphor, Herod remains a human being, but he is identified as acting in a manner that makes him as disgusting as a fox (in first-century Jewish culture). Let's review several figures of speech.

To begin, we must recognize the **metaphor**, as just suggested above was used by Jesus. When two brothers come home from a birthday party and one brother says of the other, "he was a pig at the party," he does not mean that the brother put on a costume to look like a pig, but that his brother kept on taking and eating more pieces of cake and other treats that were offered at the party. The boy was not in actuality a pig, but in some particular way his actions resembled the actions of a pig, and so the label of pig is applied to indicate the similarity. Another related common figure of speech is a **simile**. "You're like a bad luck charm. Every time I am with you something bad happens to me," says one character to another in a story. The person being addressed is actually a person, not an inanimate charm, but he is compared to a charm because of the similar effect that seems to come about with his presence. A type of comparison that is more comprehensive than either a metaphor or simile is called an **analogy**. Jesus commonly used this figure of speech: "The kingdom of heaven is like treasure hidden in a field.

When a man found it, he hid it again, and then in his joy went and sold all he had and bought that field" (Matthew 13:44). In this analogy the point is not simply that the kingdom of heaven is like a treasure. The important comparison is the behavior of the man who found the treasure. Many of Jesus' frequent analogies are called **parables** in the New Testament. A special kind of metaphor that deserves mention is called **anthropomorphism.** This is when non-human characters are spoken of as having human characteristics. One might speak of the "voice" of the wind conveying information rather than just making sound, or the moon looking down upon the Earth with eyes to contemplate what happens on Earth. Sometimes plants or animals are assigned human characteristics, such as when the trees "clap their hands" for joy (Isaiah 55:12). Frequently God is described as having human characteristics. For example, God will lift up his "hand" in anger to punish certain people, or God will rest his "feet" upon a footstool. God will set his "eyes" upon a certain person to protect him, or God will turn his "face" away from another person to reject helping him in time of trouble. Those who believe in the Christian, Jewish, or Muslim scriptures understand from those scriptures that God is not a physical creature like humans, and does not have these physical features. God is a spiritual being, and does not have a body such as humans do. Anthropomorphism is used to describe the fact that God functions in a spiritual manner to do the equivalent of the physical action when he "hears" the prayer of his people with his "ears," "sees" their needs with his "eyes," and provides them with necessary gifts from his generous "hands." This is sometimes carried to the extreme that when humans are allowed to "see" God, they see a human form. This is not a definition or revelation of the essence of God; it is a gracious condescension to the ability of the human. Believers understand this to be a special consideration for physical humans who must use eyes to perceive shape and significance. God, who is beyond any human, is assigned human features so that we can understand his action.

A different type of figure of speech is **hyperbole** (pronounced "high-PER-bowl-ly"). This is when an idea is expressed by words that extend beyond their literal meaning, and the listener knows the point is the underlying idea and not the "literal" words. When a mobster tells another mobster, "I'm going to kill you," this is ordinary speech. But when a fraternity brother tells friends, "I'm going outside for a minute; don't touch my pizza or I will kill you," everyone understands that he is not seriously intending to threaten death to any of his friends. He is using hyperbole, an extreme statement, to indicate that he does not want anyone to disturb his food while he goes away temporarily. If they do, he will be angry. But in fact he will do nothing but complain vocally to his friends if they disturb his food, or at best he will later repay with some other prank designed to make his friends annoyed. **Euphemism** is another common figure of speech where the words do not signify their simple meaning. In an old movie, when a woman at a restaurant says that she has to go powder her nose, everyone knows that this is code for making a trip to the ladies room in order to empty her bladder. Euphemisms are frequently used to avoid explicit expressions that we consider objectionable or embarrassing. Bathroom activities, sexual activities, and illegal or immoral activities are frequently covered by the use of euphemisms. For example, when a government official in a movie orders a secret agent to be killed, he will not order a "killing," but a "termination." The order to terminate does not mean to end the project, but to end the life of the agent. Sensitivity to different cultures becomes very important in interpreting euphemisms. In traditional Japanese culture it was considered very rude to say "no" to a guest. So when western businessmen tried to negotiate deals, they would offer a proposal and receive the response "yes." But they had to learn that this did not mean "yes, I agree to your offer," but it really meant "yes, that is a good offer; I will think about it, and give you a counter-offer later." In the book of Samuel we find an expression used in Hebrew, which was translated word-for-word into English by the original translators of the King James Version, but which is covered up by a euphemism today in more recent translations. The literal phrase is someone "who pisseth against a wall," meaning a male who is able to relieve his bladder while standing up (1 Samuel 25:22, 34; 1 Kings 14:10, 16:11, 21:21; 2 Kings 9:8). We sometimes encounter a modern euphemism for a similar act when we talk about those people who can "write their name in the snow." Apparently it was considered acceptable to use such language publicly by the ancient Israelites, and also by the sixteenth century English. It is not considered acceptable in polite company today, and all modern translations gloss over this with the simple translation "male" or "man." This is still considered a literal translation, since it transmits the proper meaning of the words from one language and culture into another language and culture.

Other figures of speech include such things as **irony**, **sarcasm**, and **satire**. In each of these figures the meaning of the sentence can in fact be the opposite of what the words convey. The context of the sentence makes it clear that the meaning of the sentence is not the ordinary one, but the alternate idea. In the book of Jeremiah God complains that the Judahites have not followed the Torah rules that prohibit making slaves of their fellow Israelites. During the danger of foreign invasion, the population repented and set such slaves free, but after the end of the danger, they forced the people back into slavery. God uses irony to state that he will set the Judahites "free" from his rules in a similar unhelpful way: "Therefore, this is what the LORD says: You have not obeyed me; you have not proclaimed freedom for your fellow countrymen. So I now proclaim 'freedom' for you, declares the LORD—'freedom' to fall by the sword, plague and famine" (Jer. 34:17). God sets the Judahites free from obedience to his laws, but he also explains that this means they are free from his care, and thus they are now "free" to "fall by the sword, plague and famine." In the book of Job, Job criticizes his friends who have come to help him, but whose words convey no help and no comfort: "Doubtless you are the only people who matter, and wisdom will die with you!" (Job 12:2). His words praise their great wisdom, but in the context it is clear that he is mocking their useless advice, which he then continues to expose as faulty. This can be described as either irony or sarcasm, or both.

Some other types of figures of speech can be identified as **simplifications**, **approximations**, or **generalizations**. Common proverbs often fall into these categories. They state a generalization that illustrates a basic truth, but that does not actually describe every situation. Proverbs and generalizations are not the same as the statement of scientific laws. Take the proverb "a watched pot never boils."[9] This does not mean that if someone continually stares at a pot of water which is heating over a flame, the act of watching will prevent it from ever becoming hot enough to boil. The point of the proverb is that there is a finite amount of time that will be required to transmit enough heat into the water to make it boil, and you cannot speed up that process by wishing or by watching. The actual meaning of this sentence as a proverb is that "watching a pot of water heat to boiling only causes irritation in your spirit; divert your attention to other productive activity and allow the flame to have adequate time." Proverbs are cited at special places throughout the

---

[9] Possibly from Benjamin Franklin, who attributed it to "Poor Richard" in a report he wrote for the King of France in 1785 on animal magnetism.

Christian Bible, and a special collection of ancient Israelite proverbs is collected in the book of Proverbs. An example of approximation in the Bible is when the water tank at Solomon's temple is described as being ten cubits in diameter and thirty cubits in circumference. This would seem to make the mathematical ratio of a diameter to the circumference of a circle exactly equal to three, rather than the number called *pi* (π) that we learn in math class (3.14159...). But it only takes a moment to notice that both measurements are being given to only one significant digit. The diameter is "ten cubits," that is, anywhere from 9.6 to 10.4 cubits, and the circumference is "thirty cubits," or anywhere from 29.6 to 30.4 cubits. So this is not inaccurate math, but only an approximation, or simplification, to single digits.

Another figure of speech that involves a meaning of more than the basic words is called **hendiadys** ([hen-DYE-a-dis] from the Greek meaning "one-through-two"). In this figure two items are mentioned that are usually opposites, and often extremes. The meaning conveyed is not simply these two items, but the two items and everything that would fall in between them. For example, when God creates "the heavens and the earth" (Genesis 1:1), that does not mean only sky ("heavens") and dirt ("earth"), but everything in each area and between them: all the contents of the sky, including moon and stars, and all the contents of the Earth, including all plants and animals, as well as the atmosphere, the birds, and the seas contained in or on the sky or land. Another example occurs in Jonah 3:5, when the entire community of Nineveh chooses to repent, "from the greatest to the least" (the literal Hebrew text says "from their greatest [person] until their smallest"). The mention of the highest ranking person in society and the lowest ranking person in society signifies that everyone is included, with no exceptions. Other pairs that occur frequently in hendiadys are young and old, rich and poor, male and female, or a distance "from the rising of the sun until its setting," which signifies the entire area from the most extreme east point to the most extreme west point.

Finally, one must also take account of **idioms** and **jargon** when considering figures of speech. Labeling someone as a "fair weather friend" does not refer to this person's connection to the weather, but to his behavior as your companion. He will only be around to help you when it suits him, that is, when the "weather" of all the situational conditions seems pleasant to him. All languages and cultures have idioms, and they must be recognized and interpreted appropriately. Jargon refers to specialized language used by subsets of a society. This is similar to the need to respect technical terminology which was mentioned above in connection with Rule One, but with jargon we move into a sort of slang that many small groups use. Soldiers have their own jargon, lawyers have their own jargon, engineers and salesmen each have their own jargon. When one sailor refers to another person as an "albatross," he may mean that the man is burdened with a curse that may affect others, and that man is probably better avoided. This allusion comes from a well-known 19th-century poem ("The Rime of the Ancient Mariner," by Samuel Taylor Coleridge) about a sailor and an albatross, and does not refer to any biological feature of an albatross that would function in the way a metaphor functions. The man is not similar to an albatross, but similar to the cursed state that was signified by the albatross in the poem. The label "albatross" becomes a shorthand way of conveying the idea of being burdened with a curse.

Rule Four: Historical Interpretation. The proper interpretation must be tied to the history of the time period of the origin of the scripture. It must be interpreted in the light of what is known about that history, as best we know it. As one simple example of this, take the term "north" as sometimes used in the Old Testament. All travel through ancient Israel or Palestine was north or south in direction. (By "through" I mean going in one end and out the other, not traveling around within Israel.) Palestine functioned as a narrow bridge of habitable land between the north of central Asia and the south leading to Africa, with the barriers of the Mediterranean Sea on the west and the Arabian Desert on the east. People traveled between Africa and Asia/Europe by traveling north or south through Palestine. If you went south, you arrived in Egypt. Technically, Egypt is southwest of Israel, but you traveled south through Israel and then south into the Sinai Desert, before you angled west to get to Africa. So if you were in Jerusalem and you wanted to go to Egypt, you started by going south. The other direction, north, took you into Lebanon and Syria, and after that you could angle west to Anatolia and Europe, or angle east to Mesopotamia. No one could travel across the Arabian Desert at that time, so all travel through the Middle East went around the Fertile Crescent and south through the fertile bridge of Palestine. Therefore, most of the enemy armies that attacked Israel came either from the south or the north. In one passage from the book of Jeremiah, God promises a future salvation from slavery in Assyria and Babylonia, which shall become more important than the great history of the Exodus from Egypt:

> "So then, the days are coming," declares the LORD, "when people will no longer say, 'As surely as the LORD lives, who brought the Israelites up out of Egypt,' but they will say, 'As surely as the LORD lives, who brought the descendants of Israel up out of the land of the north and out of all the countries where he had banished them.' Then they will live in their own land." (Jer. 23:7-8)

Years ago God had rescued Israel from Egypt, which was "south" of Israel. In this passage God promises Jeremiah that he will rescue Israel from Assyria, which is "north" of Israel. This usage of direction also means that when God threatens an invasion from the north, he does not mean an invasion from ancient Anatolia or the Caucasus region, the nations that were located directly north of Israel, in what we today know as Turkey or Russia, but he referred to his plans to bring the armies of Mesopotamia, either Assyria or later Babylonia, which were actually east or northeast of Israel. When God did speak of an invasion force from the east, the term "east" referred only to the countries lying inside that land bridge of Palestine, the countries immediately east of Israel or Judah—in effect, Syria, Ammon, Moab, and Edom. Invasion from the east did not mean invasion from Babylon. To get to or from Babylon, you went "north." The geographical terms have to be understood within their historical situation. Or in this example, one could say the historical terms have to be understood within their geographical situation. Geography and history are linked.

    Another major example of the need to properly interpret words with regard to history is the matter of the terms used to identify the people God chooses to work with in the Old Testament and New Testament. Should you identify the people that the Christian God works with in the Bible as Christians, or Israelites, or Jews, or Hebrews? It depends on where you are focusing in history, and it is not the same for different periods. Prior to the time of Abraham you cannot have any Israelites or any Jews, not to mention any Christians, because these names all are brought into use after Abraham. It is not clear what label one could or should use for Abraham as to his ethnicity or political and cultural homeland. He was from Ur of the Chaldeans, which many people want to link to the city of Ur in southern Mesopotamia. But the label Chaldean was a term used by Mesopotamians for people from the west, from what we call Syria, and the Chaldean language is identified with what we today call Aramaic, the language of Damascus. At some point in history enough Chaldeans moved east into Mesopotamia to greatly affect the culture, and to replace the political language of Akkadian with Aramaic. But that seems to be a thousand years after the time of Abraham. And to say that Abraham is from Ur of the Chaldeans is not necessarily to identify Abraham as a Chaldean. Abraham identifies himself as a Hebrew, not a Chaldean. But it is not clear if the label "Hebrew" is an ethnic identification, or merely a political label. The English word "Hebrew" begins with the letter "h," but the Hebrew word does not. The Hebrew word begins with a different consonant that is not used in Greek, Latin, or English. However, when the Greek writers borrowed the concept of the alphabet from the Phoenicians (who shared the alphabet of the Hebrews), they used that alphabetic symbol to represent a vowel letter, and that Greek vowel letter eventually evolved to represent the "h" sound in Latin and English. [10] Thus the English label "Hebrew" is the translation of the Hebrew word "*eberite*." This could be an ethnic label. In recent decades archaeologists have uncovered the ruins of the ancient city and empire called Ebla, which was ruled by Eberites, descendants of a certain King Eber. It may be that Abraham was descended from this family group. But the label "Hebrew/*eberite*" could also be simply a political label meaning something like "renegades, stateless outlaws, tribes without country." This is suggested by many scholars, and might fit with much of the history of that period. Abraham's grandson Jacob is identified as a "wandering Aramean" (Deuteronomy 26:5), which could support either interpretation of the label "Hebrew." Then God renames Jacob "Israel," and his descendants are known as the Israelites. Kings Saul, David, and Solomon ruled over the kingdom of Israel and the Israelites. With the death of Solomon we have the division of the kingdom; and

---

10   Compare the words hallelujah and alleluia, where the opposite effect occurred: the Hebrew "h" was lost in the Greek and the shorter form was borrowed into Latin.

one tribe, Judah, gives its name to one of the resulting countries. At that time the people living in Bethlehem were ethnically Israelite, but politically Judahite, because Israel as a nation was separate from Judah as a nation. When the Assyrian conquest brought the northern nation of Israel to an end, only the nation of Judah continued. But the people continued to identify themselves after that time as both Jews (Judahites)—ethnically and politically—and Israelites—ethnically and religiously. During the time of the New Testament the people living in Judea were both Israelite and Jewish, but as the history of the church developed, there was a need for a new label to separate the followers of Jesus from those Israelites/Jews who did not follow Jesus. Eventually they settled on the term Christian. Now, when a modern Christian reads the prophecies written in the Old Testament concerning God's plan to redeem his people, what label is used in those prophecies to identify God's people? They are not called Hebrews, but neither are they called Christians. The prophecy must use whatever was the proper cultural and political label for its particular historical situation. Therefore the prophecy will predict great blessings for either Israel or for Judah. What does that mean in application today? Does it refer now to the current citizens of the modern nation of Israel? Does it refer only to those who are ethnically, genetically Jewish? Or does it refer to Christians? Does the label "Jew" mean only those people who were descendants of the tribe named after Judah, or does it mean all of God's people at a certain time in history? When in the New Testament St. John writes negative words condemning the unbelief of the Jews, did that mean all Jews through the rest of history? Were not Peter, Andrew, Phillip, Barnabas, and others both Jews and believers in Christ? The label "Jew" in such negative New Testament statements has to mean "those who were ethnically Jewish but not Christian by faith." It cannot mean "all ethnic Jews," because many of the Jewish people were Christians by faith. We see that all of these labels need to be interpreted as to what they signified at that point in history, and then we need to carefully match up contemporary labels with those properly interpreted old labels. Our labels, our words, also are used within a specific historical situation, that of our world and culture today. All American Christians are members of the Christian church, but they are not members of the particular denomination formally called The Christian Church.

Another aspect of historical interpretation is called **historical criticism**. In historical criticism a document is searched for all possible references to any historical information, and that result is then compared to everything else we know today about the historical situation of the time in which the document was written, or purports to have been written. If the historical information taken from within the document is not in harmony with all that we know from other sources of history, then we become skeptical as to whether the document is everything it claims to be, or whether it is a document written at a later time but camouflaged to seem like it is older. Note that this test only works one way. If all the historical information taken from within a document agrees with all we know of the history of the period, that does not prove the document is genuine. It only demonstrates that the creator of the document had the same knowledge of the history of the period as we do today.

A related limitation on the use of historical criticism, particularly when used with very old documents, is that our knowledge of ancient times, and our collection of such ancient documents, are quite limited, and the document being examined might contain information that is not matched in any other source we have concerning that time period. In such a case the disharmony between the historical information from within the document and our other knowledge sometimes leaves us with an ambiguous situation. Is the document genuine and does it reveal more information to us? Or is it fraudulent, mentioning something that has no correlation? Several events from the history of archaeology demonstrate a need for caution in such cases. Let me cite one example connected with Acts 18, in which St. Paul is brought to court before the proconsul Gallio while Paul is dwelling in Corinth. In the 19th century this passage was sometimes cited as the chief evidence that the book of Acts is not historical, that Luke simply made up situations for his stories to make his material seem more credible. This proconsul Gallio was the brother of the famous classical philosopher Seneca (the father of Gallio and Seneca was also named Seneca, and is referred to as Seneca the Elder), and it was believed that from classical sources historians knew the entire career of Gallio, and that he was never proconsul of Corinth. But an inscription discovered at Delphi around AD 1900 lists Gallio as proconsul of Corinth. He served only about one year in this position. Previous to the discovery of this inscription historians knew what he did before that year, and what he did after that year, so they thought they had his entire career. As a result of this inscription, the passage that was once the most cited proof of the lack of historicity of the book of Acts is now the major fixed point around which all the other chronology of the New Testament is calculated.

Rule Five: Words Change Meaning Over Time. This is an extension of the rule of historical interpretation: one must realize that human language is a flexible tool, and it evolves with time. I mentioned above that the labels used for God's people in the Christian Bible change with time. Thus the meaning of such labels can change with time. But sometimes even more ordinary words change their meaning with time. The English word "prevent" derives from the Latin *pre-venio*, which means to come prior, to arrive before. Today the word "prevent" means to hinder, impede, or stop something from happening. In the sixteenth century the translators of the King James Version translated Psalm 88:13 to read that the poet's prayer would "prevent" God, using the older meaning. This is confusing to a modern reader. All modern translations, including most updates of the King James Version, now adjust this according to the current meaning of the English word. Now the prayer of the poet "comes before" God so that God can receive it and respond to it. In a similar way the English word "meat" once could be used to mean "food" and not necessarily animal meat. Also the word "corn" did not refer to the modern plant that was native to North America and not known in ancient Israel. The King James Version translators used the word "corn" to refer to grain plants such as wheat or barley. One must be even more careful with the connotations of words beyond their basic definitions. Jesus calls King Herod a "fox" in Luke 13:32. In America today that might be taken as a compliment, for we have developed the idea from children's books of the sly, crafty fox. But in the world of the New Testament no such association existed. It might communicate better if we were to translate Jesus' reference to King Herod as a "jackal" to bring out the insult intended.

Rule Six: The Intention of the Original Author. This rule involves something that has been mentioned in the above rules, but not brought out explicitly. Proper interpretation of a text requires seeking to determine the intention of the original author. This principle might seem obvious: what else could it mean when you read what someone else wrote, but to understand what the writer meant? But research on the use of literature, including religious literature, has made it clear that this is not always as simple as it sounds. Particularly when one deals with literature considered to be important within a certain culture, such as sacred scripture when one religion dominates the culture, parts of the literature tend to be used to undergird certain doctrines within the culture. The *interpretation* of a particular passage may take on a life of its own, and in time the passage can come to have a meaning for the members of the culture that may be quite different, even the opposite, of what the original author intended. For example, the story of Noah and his sons after the flood came to be used in America as a Biblical passage justifying the slavery of African peoples. After Noah became drunk and disgraced himself, he placed a curse on one of his grandsons, Canaan, the son of Ham (Genesis 9). (Perhaps Noah felt that since his son Ham had dishonored him, he would dishonor Ham's son in revenge.) Scholarship of the 19th century associated Noah's sons with the division of the different races found in different continents; Ham was interpreted to be the ancestor of the African peoples. Noah declared that Canaan would become a slave to his uncles. In the United States, this interpretation was taught as a religious justification for the maintenance of slavery of the African peoples. Modern scholarship rejects any such interpretation. The curse is addressed to Canaan, not Ham, and the descendants of Canaan are considered to be Middle Eastern, not African. Therefore this story might have something to do with justifying the conquest of the Canaanites by the Israelites, but it does not really speak about Africans at all. In addition, it is probable that this act of Noah was not a God-pleasing act; like all the other leading characters in the Bible, even Noah was not perfect and he also sinned, doing things that displeased God (in this story, getting drunk, acting shamefully, and then blaming his son for his shame). The interpretation that Noah's curse signified God's design for African slavery is completely foreign to the original intention of this story. But the use of this story in later European and American culture came to have its own meaning. It sometimes requires great intellectual effort and significant courage to question an interpretation that has become accepted in one's culture. Yet it is necessary for intellectual honesty to ask the question, would the original writer and his readers have agreed with the interpretation of the text that some part of current culture has expounded?

The work of literary criticism in recent decades has highlighted the fact that every reader brings to the interpretation of any text his own set of prejudices. When a person reads a text, that person's internal understanding interacts with the information from the text so as to produce ideas that he can work with in his mind. Sometimes the experiences of the writer and the reader are so different that the reader cannot view the text in the same way as the original writer. In the late 1800s after the American Civil War, the author Mark Twain wrote the novel *Huckleberry Finn*, in which he subtly but strongly criticized the existence of racial prejudice in America. The book generated controversy at its appearance for the language and attitudes used to depict the characters in the book, and it still generates controversy today when proposed for use in high school classes. It describes a world in which black slavery was assumed, and its dialog speaks in the racist phrases of that world. The dialog and the toleration of the social situation of that time are considered so wrong today that just to read the language used in such a book is considered offensive by some people. It requires education and effort to distinguish the world which the book describes from the world of today. It requires effort to try to understand the situation and message of the original author. In the case of this book, we are looking at a piece of writing within the same culture (the United States) after only a little more than one hundred years of change, concerning which piece of literature the author is strongly praised for the clarity and power of his presentation of his point of view. And yet many contemporary people find it difficult to properly appreciate this book.

This problem of taking the different worlds of author and reader into account becomes even more complex when considering other types of literature. Poetry, hymns, fables, fairy tales, and children's stories often are written in a manner that does not state a direct, simple point, but which seeks to create an impression that stimulates either feeling or thinking on the part of the reader. This difference in interpretation can be illustrated by the old joke about two British shoe salesmen who traveled to India. One wrote home and asked for funds to return to England, explaining that it was hopeless to make any money because "nobody wears shoes here." But the other wrote home excitedly asking for more shoes to be sent to him, for he explained: "Nobody wears shoes here!" What one perceived as a hopeless marketing situation, the other perceived as a wide-open marketing opportunity. Another example of how people can view a story from their own perspective and find a message different from that expected by another interpreter is found in connection with the novel *The Grapes of Wrath*. John Steinbeck wrote this book about the problems of poor Americans during the Great Depression and the years of the Dust Bowl, when many farmers in the Great Plains had to give up and migrate to California, to seek work in California agriculture. This story was made into a movie in 1940. Joseph Stalin imported many copies of this movie into the Soviet Union and had them shown to the Soviet people because he thought this movie illustrated the downtrodden and desperate situation of the poor lower classes in the American capitalist economic system. He felt this would be good propaganda supporting the communist system. But when the Soviet farmers viewed the film, what they noted was that in America, even the poor had trucks to drive to California instead of having to walk. The details of the content of the story communicated a different message to the viewers than the one intended by Stalin.

In recent decades some scholars have promoted the idea that a written text has its own existence distinct from and separate from the intention of the author, and that when it interacts with any reader, the result of the interaction is different for each reader. In other words, every reader finds his own meaning from the reading of any text. There is both truth and error in this idea. Certainly, because every human is unique, each person is affected differently by reading a piece of literature. What one person finds boring, another person describes as the most interesting thing he has ever read. But it is not correct to push this idea to the extreme that suggests that a piece of literature has no meaning in itself, but only the meaning that is created in the mind of the reader. To go to this extreme is to deny the existence of the writer. A piece of literature only comes into existence when it is written by a human author. The author used human language to attempt to communicate his thoughts to someone else. There are two parts to the response to this communication that is carried out by the reader or listener. There is the mental part, where the reader attempts to understand what is being said by the writer. In this part the reader seeks to understand the thought propounded by the other person. The other part is the reaction of the reader. This reaction may be inspired by the ideas within the communication, or it may be a reaction to various aspects of the manner of communication. For example, when one person talks to another, the listener may react more to the tone of voice of the speaker than to the simple words. The words may convey a neutral question, such as "Where have you been?", but the tone of the speaker may convey great hostility or threat, and the listener may choose to react to the tone rather than to the words. This can happen even when the speaker does not intend hostility. A wife can become very angry at the behavior of her children, and just at that moment her husband can arrive home. When she snaps out the question "Where have you been?" her tone contains the frustration and anger built up by her children. But the husband hears the hostile tone and takes offense at this non-welcoming address. It requires a bit of extra thought on the part of the husband to analyze the situation in this way: "I am not arriving home late; I have done nothing wrong to make my wife upset; this emotional tone on her part is improper in this situation and not normal for her; there must be something else going on that has made her upset, and she is probably not mad at me, but mad at something else, and she is trying to redirect her attention to me." That husband would reply calmly, rather than defensively, and would seek to help his mate in her situation of stress. That is an example of sorting the two parts of the response to a communication.

    The example above involved oral communication, but the same idea is true of reading written communication. A piece of written literature has the ability to convey much by means of the literature's phrasing, its use of adjectives and adverbs, and its arrangement of its information. In poetry or hymns the word order can be changed from that of a normal sentence so as to place extra emphasis on certain words, often by moving them to either the beginning or ending of a line. "To the store I went!" conveys more information than the mere "I went to the store." When a reader takes on the task of reading, that is, interpreting, such text, he takes on the burden of both reacting to the literature, and seeking to understand what was in the thought of the original writer. It is quite fair to give attention to the intellectual or emotional reaction that is generated in the reader by a piece of literature. But it is necessary to distinguish that reaction on the part of the reader from the intent of the writer. The original writer had his own thoughts and feelings that he was attempting to convey. The reader does not do justice to a piece of literature unless he recognizes that and seeks to perceive the intention of the original writer. It is true that in some cases we are so removed in time, culture, and situation from the original writer that we cannot always determine what he meant to convey. In one sense this is because we lack enough of a historical and cultural context to fully interpret the author's meaning. But the fact that we have such limitations cannot be used as an excuse to forgo the attempt to interpret the intention of the original writer.

One of the tools that can be used to help determine the intention of the original writer is to try to analyze the **original intended audience**. A writer always has some target audience in mind when he writes out his thoughts. When a reader takes account of this intended target audience, this often enlarges the context in which the piece of literature should be understood. If the writer is intending to address a highly educated group about a technical subject, that will be reflected in the choice of words and jargon that are used. If the writer is intending to address a popular audience, the same words might require a different interpretation that fit the usage among the popular audience, rather than the technical meanings used by specialists. For example, the words "stress" and "strain" have very specific meanings in engineering, but a popular audience may understand these words in an opposite sense than an engineer would; the writer needs to take account of the likely interpretation by his intended audience. As another example, a religious writer uses a different style and selects different topics when he is writing to a group of believers than when he is writing to a group of non-believers.

Rule Seven: The Purpose of the Text. The seventh rule extends the sixth rule (seeking the intention of the original writer) into a broader picture of how we understand what we are reading. We must remember to keep in mind the purpose of the text. The revelation of truth conveyed in the production of sacred literature is about the supernatural. It is about religion, or about a god, or the spirits, or the other unseen world—the supernatural world. You cannot take a piece of literature that is intended for one purpose and force it to serve as the communication of information for a different purpose.

For example, if an author wrote a political history about European leaders, he might mention certain European musical composers with whom the leaders interacted, such as a certain king provided financial sponsorship to a certain composer, or a certain symphony was written in honor of a certain king. But this information cannot be taken as definitive information about the history of European composers. Because it is incidental, the writer of the political history may choose to insert that information at any point he feels appropriate. It need not be in a proper chronological order. He may choose to include information about the relationship of a composer and a political leader at a point when he is beginning the discussion of that leader, even though the interaction with the composer actually happened years after the beginning of the political career of that leader. The information about the composer is "true," but we may not presume that the writer is intending to tell us details about the date of the life of the composer. A political history of Europe is not a musical history of Europe.

A cookbook is not a chemistry book. A history book is not a science book, nor an economics textbook, nor a psychology textbook. A book on politics is not a book on sociology or social structure, nor a book on economic policy. A book on animal husbandry is not a book on biology, despite the close relationship between these two topics. There is a great deal of information that belongs in any modern text on biology that would be omitted from a text on animal husbandry. A book on religious worship is not the same as a book on music, despite the important role that music often plays in religious worship. No one can write everything into one book. That is too much information. Every book, every piece of literature is intended to focus on its particular topic. It contains information from the writer to the reader on that topic, and it does not pretend to convey any fullness of information about other topics. For those religions that focus on sacred scripture as revelation from the divine world, the topic is God or religion. When the supernaturally inspired author wrote his information, he was trying to explain something about God. He was not writing a history of the world; he was writing only about the history of human interaction with God. He was not writing a science book; he was writing only about the relationship of God to nature. He was not writing a text on sociology or human mating, sex, and marriage; he was writing about what God had to say about such topics. If you want to know all about any of these topics, you cannot use sacred literature as your sole source. It does not intend to explain all you need to know. The Christian Bible has much to say about the use and misuse of money, for example, but this is not an adequate source to explain economic policy, or even to fully teach about family budgeting. The writer of sacred literature assumes the reader has access to other information and can learn more from that other information about secular (worldly, natural) topics. It is not the purpose of sacred literature to substitute for that information. The purpose of the revelation is to relate what God has to do with some particular topic.

Phrased this way, it probably seems obvious that the purpose of sacred literature of revelation is not to substitute for other types of literature and other sources of information. But this mistake is commonly made in many religions. One does not have to search very hard before one finds people arguing that the Bible contains all we need to know about family structure, or the Bible contains all we need to know about politics and economics, or the Quran contains all we need to know about social justice and criminal law. The respect for the authority of the sacred literature leads some to use sections of the literature as authoritative in other areas of knowledge. Thus the Bible or the Quran or other sacred literature come to be viewed as books about history, or books about natural science, or books about human psychology, or books about social structure. These views are not valid. The religious literature is about God, not about geology or mathematics or economics or family structure. The information about how God might connect to those topics is important, but the focus is on the information about God, not simply on those topics. The information about such topics is incidental. It does not intend to be complete. This leads to a simple guideline: **When you ask a scientific question about the natural world, you use science to obtain an answer. When you ask a religious question about God or the supernatural world, you use religion to obtain an answer.** You do not mix the two games, or mix the rules of the two games. Science cannot tell you anything about the realm of the supernatural, or about God, by the definition of what science is today. And religious literature does not explain the natural world to you, by recognition of what this particular literature is and what its purpose is.

Let us look at a specific example. If you want to know what is the Christian God's design and plan for normal human marriage, you consult the Bible. You will learn about God's design in the creation of Adam and Eve, and about God's dislike of divorce. But if you want to know about sexuality, what is sexual desire, how does it work, how much is genetic and how much is cultural influence, how does sexual biology work in men and women, what is the process of sexual maturing and later aging in humans, how does fertility and reproduction work, how does contraception work, what are the proper social roles for men and women, what are the possible methods for raising and educating children, how does the process of wooing/courting/seeking a mate work in different cultures (are the rules for dating the same during wartime as peacetime?), when is divorce an acceptable option in a sinful world and what are the best rules for post-marriage relationships, all of those questions require something other than the Christian Bible (or the Quran for Muslims). The Bible speaks to the topic of sexuality, but it does not pretend to fully explain the topic. If you try to form your understanding of proper sexuality just from the information in the Bible or in the

Quran, you are going to have a very hard time trying to clarify questions about monogamy and polygamy, the dating process, or proper standards of public dress in connection with sexuality. The Bible does not speak clearly about such things, because it was not designed and not intended to do so. The Bible and the Quran reveal God's opinion about certain types of behavior in connection to sexuality in certain social situations. From those particular examples the believing reader must develop an extended understanding of how to apply God's will to many other situations involving sexuality. But things can become quite complex, and many situations involve more than simply sexuality. As one modern example of this complexity, take the case of older adults in senior homes. Recent American economic policy entitled senior citizens to certain financial support from the government (Social Security) depending on their previous marital relationships. If they remarried, those previous relationships, and all financial support based on them, were canceled. So we had widows and widowers living together in nursing homes who desired to form new marital relationships, but who could not afford to do so due to economic policies based on beliefs about how society should support marital relationships. Neither the Bible nor the Quran speak to such situations; the cultural setting today is completely beyond anything experienced or contemplated by people in the time those scriptures were written (the average lifespan was shorter, there were no nursing homes, there was nothing like Social Security payments).[11] In order to try to determine God's will for such modern situations, you need to *bring to* the interpretation of sacred scripture much information about both human sexuality and economics that is not "revealed" in scripture, and then you must perform a significant amount of

---

[11] There has been some recent modification of the Social Security rules to try to improve such situations.

logical reasoning in order to *move from* the comments within sacred literature that indicate God's will.

So when you read and interpret sacred religious literature, you need to keep its focus in mind. You need to keep clear what it is, and what it is intended to communicate; and what it is not, and what it is not intended to explain. You must bring to the interpretation of sacred literature all that you know from other ordinary human sources: This means you must not only bring all you know from other sources about grammar and history, but you must also bring all you know about other secular subjects, such as science and psychology. To rephrase this, **you use the proper processes of scientific or historical research to learn all you can about a subject from the natural or worldly side, and then you bring that information with you when you consult the sacred literature for what God might say about that subject.** You do not start with the sacred literature to research a secular subject. It is not the purpose of the sacred literature to be a primary textbook on any secular subject, so it cannot and does not attempt to provide comprehensive information on any secular subject. You must learn all you can about the secular subject from secular means, and then you consult the sacred literature to learn what God might have to say about the topic.

To give one example of this, the Old Testament is not a history book. It contains a great deal of information about the kings and prophets, including lengths of the reigns of the kings, but it is hardly a complete history book. Often the book of Kings states that other material about a certain king was available in other official books, which books unfortunately have not survived through history to our time. The Old Testament will skip over decades or in some places over centuries in its summary of history. This is not because nothing important in human history happened in those time periods, but because the writer has nothing significant to say about those time periods and their events in connection with God. In some cases, such as the Black Obelisk of Shalmaneser III,[12] we have learned significant information about Biblical characters and events that is not mentioned in the Old Testament. The omission of this information from the Old Testament does not mean that the information is unimportant, but only that the writer had nothing to say about that information in connection with his teaching about God.

A second example of this principle is the discussion of the age of the planet Earth. This is a scientific question, so you begin by using all the resources of modern science to learn all you can about the age and evolving status of the Earth. Then you turn to sacred scripture to see what God may have revealed about his connection with the Earth. You do not start with scripture in an attempt to gain "historical" information, in this case the date of creation, and then bring that information to the discussion of science. The purpose of something like the Bible is to reveal supernatural information, such as the matter of God's creative activity. The Bible is not a book of science or history. First you determine from scientific evidence such as radioactive dating that the Earth is about 4.5 billion years old. Then you seek to learn from the Bible the significance of God's creative action and God's

---

12   This stone monument contains the only pictorial representation of any king of Israel or Judah that has been found by archaeologists. It contains engraved pictures and a brief text that informs us that King Jehu of Israel accepted vassalage to Assyrian Emperor Shalmaneser III, which was apparently a complete reversal from the previous position of Israel in international politics. The Old Testament tells us about King Jehu's policy in regard to religion and his devotion to the Biblical God, but it tells us very little about Jehu's international relations, especially in connection with Assyria. This information from archaeology allows us to interpret what the Old Testament does say in a way that gives us insight into how Jehu negotiated a compromise between religion and politics.

intentions for the design of the planet and for the care of the planet. More will be said about this in a later chapter.

Rule Eight: Textual Criticism. Textual criticism is the technique of evaluating different copies of a piece of literature that have survived through history, when these different copies have slight differences in their words, in the attempt to determine the history and reason for these variations. Sometimes scholars of textual criticism are able to produce a probable history of the creation of the textual variations that we can accept with great confidence, and sometimes the available data is such that no one can confidently choose between two or more variations as to which was most likely the original text. Research in this area requires scholars to take account of a multitude of factors. They have to study the history of the process of copying texts, the history of the writing systems, the history of theological and political disputes in different parts of the ancient world, the history of the spread of the texts to different geographical areas, and the relationships of ancient scholars and officials as to who would be influenced by whom. They have to apply various principles that they have worked out to help explain what kinds of mistakes are most likely and least likely when copying texts. They also have to examine ancient translations of the text into other languages in order to deduce which version the ancient translators had available at different points in history.

In the religious use of sacred literature, it is often not wise to draw any significant conclusions from a part of a text that is uncertain due to the existence of such textual variations. The rule of context mentioned above comes into play again in such instances. If we are uncertain what a particular line of text in the sacred literature originally said, it is best to leave the uncertainty stand, rather than force a decision one way or another. It is possible that future discoveries from archaeological work might provide data that can resolve such uncertainties. In the meantime, the clear statement of other passages of the sacred literature should guide the understanding of that particular topic. But when the work of textual criticism has produced results that would negate the reading of one text and confirm the reading of a different text as very highly probable, it is necessary for the reader to accept this information and work with the consensus text, rather than try to defend the supposed superiority of some familiar or favorite text. In the Christian Bible, some people today try to insist that the readings of the King James translation of the Bible should be followed in spite of the evidence of textual criticism. Some examples of passages omitted from modern translations include 1 John 5:7b-8a, Mark 9:44 and 46, and Mark 11:26. The English scholars of the sixteenth century worked with the best information they had at that time. Today we have better information, and good interpretation requires that we use it.

Rule Nine: Humility. The reader of sacred scripture ought to maintain a humble attitude that places his understanding or his interpretation under the authority of the revelation from the supernatural world. The reader has not been to the supernatural world, and the reader has no direct knowledge of any part of the supernatural world. God, or whatever else a particular religion believes operates the supernatural (spirits, fate, etc.), has chosen to reach into the world that we can sense in order to communicate with us. We must be very careful that we do not place ourselves in the position of controlling what it is that God can say. The supernatural revelation stands over the interpreter. This calls for humility in several aspects. Sometimes we cannot decide what the text says. Perhaps there is the existence of different variations of the text that conflict in what they say, and the existence of these variations leaves us uncertain of what the original meant. Perhaps there is a problem with the grammar that leaves us uncertain how to understand the sentence. Perhaps there is a word (or more than one word) in the text that we do not understand. We do not have the liberty to create a meaning or determine a meaning where the normal rules of human language do not produce a clear meaning. Any such action has the effect of placing the interpreter over the text, and nullifying the ultimate authority of the sacred literature. This would take us back into the situation of the other types of supernatural intervention, which depend upon some authorized interpreter. Then the sacred text is reduced to mysterious symbols, and we are at the mercy of whoever claims such interpretive authority. The existence of sacred literature as the intervention of the supernatural world into our world requires that the literature be allowed to function as literature. It is intended to be read by any capable reader. If something obscures the reading and comprehension of the text, then the reader cannot determine what it means. No one has the right to determine for the reader what the text says in such a case. Everyone must accept the uncertainty of such a portion of the text. Sometimes this results in different religious ideas being preferred by different readers. If neither a specific text nor the total context of all the sacred literature speaks so as to clearly decide such a question in one way or another, then all parties must leave the question open. Humility is required of the human readers. No one can insist that the text must mean what it does not clearly say.

The Role of Human Reason in Supernatural Interpretation

Before I move on to discussing the tenth and last rule, I think that this is an appropriate place to provide some further discussion on the **role of reason in the interpretation of sacred literature.** In the discussion of the rules of science in an earlier chapter, it was mentioned that the relationship between science and religion has often been discussed in terms of the relationship between reason and faith. When you phrase the issue that way, it suggests that there is a conflict between the use of human reason and the following of religious faith. This aspect of the relationship between science and religion, namely reason and faith, is important and useful when you are investigating certain philosophical questions as to *how* we seek to decide what we think is truth. But when you engage in that investigation of *how* to seek answers, you are neither playing the game of science, nor the game of religion. You are playing the game of philosophy. When you play the game of philosophy, you need to follow the rules of that game of human research. As mentioned before, the confusion of different games has impeded proper understanding of many things concerning the relationship between science and religion. In this book the focus is on sorting out the different rules of the games of science and religion, and distinguishing both of these from the game of philosophy. In sorting out these rules it is *not* helpful to set up a contrast between reason and faith. That simple contrast seems to suggest that you do not use human reason when engaged in religion. Nothing could be less accurate. If you say that chess is a game of logical strategy and is different from checkers, that does not mean that the game of checkers does not use logical strategy, even though chess and checkers are different games. In both the games of science and religion we are using human reason to determine what we hold to be true. It is not the case that we use reason to interpret the natural world and to deduce the scientific processes by which the natural world functions, and then do not use reason in religion. Just as one uses reason and debate in researching scientific truth, one also uses reason and debate in researching religious truth. All of the previous discussion of the rules for interpreting sacred literature engages the function of human reasoning to determine what the text says. Stepping back for a moment to the larger possible sources of supernatural revelation besides sacred literature, the interpretation of supernatural symbols or the interpretation of supernatural intervention by events in history also require human reason to determine the meaning of the supernatural communication. In fact, in a real way both the games of science and religion have moved beyond the old philosophical question of reason versus faith. In the case of science the shift to the importance of experimental scientific research represents a critique of deducing truth in science simply by careful reasoning. It is now recognized that reasoning alone can easily overlook certain facts of natural phenomena. In addition, there are often other aspects of natural phenomena that are not known until experiments are performed, and these new pieces of data are "accidentally" discovered. Science demands careful reasoning, but it also demands careful experimentation. In the area of religion, reason is demanded along with faith. The modern focus on sacred literature as the primary source of information revealed from the supernatural world demands the use of human reason to interpret the literature. Just as modern science is not content to follow what some scholar thinks, but demands experimental evidence, so in modern religion it is not enough to hear what some theologian thinks; the believers demand the demonstration of the accuracy of his thinking by reference to some clear text of revelation.

In the field of Christian theology, the use of human reason in religious interpretation is often discussed by distinguishing two categories of the use of reason: **the *ministerial* use of reason and the *magisterial* use of reason**. The ministerial use of reason refers to the use of reason in a ministering or helping sense. In this sense a person uses human reason to its fullest extent in interpreting the phenomena that come into his awareness. You interpret what you sense in nature, and you interpret what you read in sacred scripture. You do not just react to phenomena, you logically reason about the significance of the phenomena.

In my house we have cats as pets. Two of those cats react quite differently to the sound of thunder outside our house. One cat basically ignores this sound. He understands that he is inside the house, safe from rain and all danger. The second cat does not understand this. When he hears thunder, he immediately gets a fearful look and seeks shelter. For this cat, shelter consists of hiding under some low covered item, perhaps under a bed or a dresser. You cannot tell the cat that everything is okay, that he is indoors, and he is safe. He simply believes he has to seek shelter under some covered item. However, if you take the first cat, who ignores the sound of thunder when indoors, and carry him outside to sit under the porch roof to watch the rain, the experience of outdoor thunder makes him panic and he seeks to flee to some shelter. For him, being indoors or outdoors makes all the difference. Each of these cats has his own logical thinking about the phenomenon of thunder. But you cannot discuss the issue with either cat because they do not communicate with human language. With a human child, you could discuss the details of thunder and lightning and the relative shelter of being in the house or under the porch roof. Using language, you could engage the human child in a reasoning process, which could change his understanding of the significance of thunder, and guide his future reaction to thunder. In addition to providing the child with a reasoned discussion of the matter of physical shelter from lightning, you could also exercise this use of reason in connection with the religious implications of lightning. You could discuss the scripture passages that comment on God's control of rain and thunder, and with those texts you could convince the child that the fact that lightning struck the tree in the neighbor's yard does not mean that God is angry at your neighbor. You could point to scientific information about electrical potentials and the height of the tree, and you could point to texts that emphasize the Christian gospel and God's assurance of love for your strong-in-

Christian-faith neighbor. All of this is the ministerial use of human reason.

The other category for the use of reason is the magisterial use of reason. In this use the process of human reasoning is shifted to a "magisterial" role, the role of a magistrate or governing official. In this sense logical reasoning is assigned a level of value that makes it determinant over other aspects of information. For example, suppose there is a crime committed at a home in a certain neighborhood, and the surveillance camera only shows one person walking down the street that night. A simple logical reasoning concludes that the person pictured in the camera recording must have been the one who committed the crime. A more careful reasoning would raise the question whether there might have been someone else who did not walk on the street in view of the camera, but who crept through the alley, or hid in the vegetation. But a magisterial use of reason would demand that the simple reasoning process gave the answer: there was a crime, and only one person was spotted in the area, therefore that person had to have committed the crime. This magisterial use of reason, when a process of logical thinking is allowed to define the situation and limit the possible conclusions, is a problem for both science and religion. In the area of science, such a process of reasoning must be confronted with experimental evidence. For example, we know that the force of gravity pulls harder on heavier objects than on lighter ones. Reasoning suggests that this must mean that heavier objects would fall faster than lighter objects. But Galileo is famous for disputing this long-held ancient assumption. By experiment, the story goes, he demonstrated that balls of equal size fell at the same speed, even though they did not have the same weight. The process of acceleration by gravity is not as simple as some types of logic might desire. In the area of religion, a magisterial use of reason is often used to deny the possibility of miracles. No one who dies ever comes back to life again, no one is cured of cancer or blindness by simple words, no river ever splits apart to let someone walk across to the other shore. These things are not part of the regular process of natural events, and no experiment can replicate such actions. Therefore they can never happen, the magisterial use of reason tells us. If the sacred text says that any of these things happened, the text is wrong, and must be rejected. The text cannot be "truth" if it tells us something that conflicts with what "reason" says is possible.

There are two problems with this controlling use of mere reason. One is that coincidences and apparent miracles do happen in history. From time to time humans who are ill do experience complete healing, to the astonishment of doctors. People who are pronounced dead later wake up in the hospital or in the morgue. And miracles occur in connection with natural phenomena from time to time, such as breaks in storms, or diversions of floodwaters, or a child being picked up, carried, and deposited by a tornado without any harm. These similar events are dismissed as the mere result of chance, perhaps involving factors which we simply do not yet understand (in the case of miraculous healings). These "chance occurrences" are not allowed to play any part in the discussion of whether a miracle reported in sacred scripture could be true. Suppose you were lost and in a desperate situation, and you prayed to God for help. Then out of nowhere some help appeared that led you to a safe situation. This personal experience would not be allowed to have any spiritual meaning by some people because the magisterial use of reason has already concluded that the act of prayer does not work to contact any higher power. This magisterial use would require that your personal experience be reinterpreted in the light of what this controlling use of reason dictates must be the limits of truth.

The second problem with this dictate of the magisterial use of reason denying miracles is that it is itself illogical. The sacred text reports that it is an account of supernatural action. The magisterial use of reason says there is no supernatural action. Therefore the text is not allowed to speak. This is equivalent to the king saying that his subjects love him and would never rebel, no matter how much he taxes them and abuses them. When the rebellion comes, he is confronted with something that he does not believe can be happening, and as a result he is unable to properly respond to it. The logic is faulty. To say that the supernatural does not exist and cannot act is not to establish that it neither exists nor acts. As Galileo proved by his experiment that merely reasoning about gravity was inadequate, merely reasoning about the supernatural is also inadequate. The religious experience of humans suggests over and over again that there is something more to the world than the domain of normal natural processes. According to the testimony of people, there are miracles, there are coincidences, there are dreams, there are spirits, there are demons, there are voices, there are healings that are not normal. They are, according to the testimony of those who experienced such things, beyond the natural. The fact that in some cases unusual stories can be explained scientifically (such as when it is found that a house thought to be haunted was not really haunted, but someone was faking the voices and the mysterious opening and closing of doors) does not mean that all such accounts are false. The old story about the boy who cried "wolf" in fraud so often that he became ignored by the population depends on the detail that one day the real wolf came and no one responded to the boy's cry, to his detriment. To say there is no wolf is not to establish that there is no wolf. And to say there is no supernatural is not to establish there is no supernatural action.

Therefore we see that we must be very careful how we phrase the question. "Reason versus faith" is a valid topic in the game of philosophy. But in the games of science or religion, faith is required to follow the evidence that comes from proper research. Science demands the use of reason and experimentation to establish its facts and scientific laws. Religion demands the use of reason and revelation to establish its facts and religious truths. The human mind is required to read the sacred text and observe the revelation, and process the data. The human mind is required to properly interpret the data that the supernatural world chooses to reveal, and the supernatural world considers the human in error and at fault if he does the interpretation incorrectly. The ministerial use of reason is required in the game of religion, just as it is in the game of science. The difference is in the kind of data that is being interpreted. Science interprets the natural world. Religion interprets the supernatural world. Both science and religion have to guard against a magisterial use of reason, in which a logical chain is allowed to have authority over the evidence of either science or religion.

Rule Ten: Consistency with the Central Goal of the Religion; or Christ-Centered Interpretation. There is one last rule to be offered for interpreting religious literature in the game of religion. Phrased in a general way, this rule states that **all religious interpretation must be consistent with the central goal of the religion**. It is thus simply an extension of the rules mentioned above about keeping the purpose of the text in focus and seeking the author's intention, along with keeping the entire context of the sacred revelation in mind. Some religions might not condense to one central principle as easily as can be done with Christianity. Nevertheless, even if a particular religion cannot be focused to a central principle, it is always worth asking the question to what extent the interpretation of that religion's sacred texts is connected to its key principles.

For Christianity, this rule can be focused very easily: **all proper interpretation of the Christian Bible must be Christ-centered**. For Christianity, the heart of the Christian revelation is the gospel of Christ Jesus. The Christian understanding of the Bible is that all of God's revelation is part of his plan to fix the great injury caused by human sin and restore the relationship of love between God and his humans through his great act in sacrificing Christ Jesus. Everything serves this greater purpose. Thus some parts of the supernatural revelation in Christian theology serve to help expose the problem of human sinfulness, and some parts serve to help clarify the saving grace of God in Christ. Therefore all proper Christian interpretation of the Bible must be Christ-centered: it must be connected to this central purpose and help support it. If the interpretation of some part of the Christian Bible has nothing to do with Christ, or seems to be in contradiction to the Gospel, or has nothing to do with this Gospel, then that interpretation is suspect. It may be incorrect. It may be reading some meaning into the text that does not represent the intention nor the revelation of God. Thus the rule about humility and caution comes into play again.

This has been a long discussion of the rules of interpretation of sacred literature in the game of religion, so it is appropriate to end this chapter by giving a quick summary of the rules. Religion as defined in the modern age is the study of the supernatural world, in contrast to the study of the natural world. Therefore religion is the study of supernatural revelation, for that is the only way of knowing anything about the supernatural world. In most religions in the modern world, that comes down to the interpretation of the written revelation of their sacred scriptures. The interpretation of the sacred scriptures requires the following:

- 1) The sacred scriptures should be read as ordinary human language, that a normal literal interpretation be performed that is subject to the inspection and approval of any normal reader.
- 2) The interpretation must respect the total context of the sacred scripture, interpreting each part in a manner consistent with the whole.
- 3) A literal interpretation is one that takes full account of normal figures of speech used in human language.
- 4) A literal interpretation also respects the historical aspects of the text, both in what it says and in terms of the circumstances of its composition.
- 5) This literal interpretation includes an acknowledgment that in human language words and phrases can change in meaning over time, and the proper interpretation respects the time of origin of the text.
- 6) The goal of a literal interpretation is to find the intention of the original author, and research toward this goal can sometimes be helped by careful consideration of the original target audience.
- 7) A proper literal interpretation must keep in mind the purpose of the revelation, that this is communication from the supernatural world about God and the supernatural, and it must not be turned into some other kind of literature with some other purpose.
- 8) The need for good textual criticism must be recognized.
- 9) The reader must maintain an attitude of respect and humility before the text. The reader cannot

dictate or control what the text is allowed to say, but must use the full process of human reasoning to determine what the author intended to say as the revelation from the supernatural world.

- 10) For Christian literature, all interpretation must be Christ-centered, that is, it must function properly within the revealed central goal of the revelation. Other religions may have a similar central principle guiding their interpretation.

Chapter Seven: The Problem with Intelligent Design

Intelligent Design is an idea proposed in recent decades that seeks to take account of new information gathered from the sciences of biology and physics in order to question the prevailing philosophical framework that natural evolution comes about by merely mechanical and statistical processes.[13] The Intelligent Design movement holds that the intricate complexity of many features that we observe in nature, notably in biology but also in the details of astronomical cosmology and in the details of atomic physics, points toward the supposition that these features were designed by some intelligent agent, rather than happened by chance. In cosmology, there are a number of features, including the strength of gravity and other atomic forces, the size of our Milky Way galaxy, the position of our planet relative to our sun, and the composition of elements making up our planet, that have to be exactly the way they are, or life as we know it would not exist.[14] Making any change in these features would result in a planet and solar system that would not permit the chemical functioning of life as we know it. In biology, the intricate and

---

13 For further information see Michael Behe (1996), *Darwin's Black Box*, New York: Free Press; or William Dembski (1998), "Intelligent Design as a Theory of Information," at http://www.arn.org/docs/dembski/; or James B. Miller, editor (2001), *An Evolving Dialogue: Theological and Scientific Perspectives on Evolution*, Harrisburg, PA.: Trinity Press International.

14 For further information see Charles Edward White, "God by the Numbers: Math and the Theology of Origins," and Howard J. Van Till, "What Good is Stardust?: The Remarkably Equipped Universe," chapters 5 and 6 in *The Origins Debate: Evangelical Perspectives on Creation, Evolution, and Intelligent Design* (Ver. 1.0), from the Editors of *Christianity Today* in the series *Christianity Today Essentials*, Carol Stream, IL: *Christianity Today* (2012), an electronic book at www.ChristianityToday.com.

complex chemical processes within cells, and the types of apparatus found within microbial creatures function together in such a specific manner that it is very difficult to imagine a process of evolution that created such a structure. So many pieces depend on a precise interaction with other pieces that it is hard to imagine a situation in which these pieces could have existed and functioned without the other pieces. But if these complex pieces could not exist without the other complex pieces, then it is hard to imagine a process where the mere chance combination of chemicals resulted in the creation of these pieces such that they could suddenly find each other and interact the way they do to power the processes of life within cells today. Using an analogy as an example, suppose you were a visitor from outer space coming to our planet, and you found our Global Positioning System (GPS) satellites in orbit broadcasting their radio signals, and then on Earth you found a GPS receiver with a map display, you would know that such a system was not a collection of wires and electronic parts that came together by chance, but it had to be a system designed by someone to accomplish the specific purpose of providing location information. The Intelligent Design movement suggests that scholars need to face up to the implications of similar data in biology and cosmology: that there is some intelligent mind that designed the creation of the world that we experience.

    The problem with the Intelligent Design movement is that some Christians seize on this idea to support their belief in religion. They reason that if the scientific evidence points toward the existence of a designer, then that designer corresponds to God the creator, and thus the data from science now support the idea that there must be a god who designed and controls the universe. In the opinion of these people this means that anyone who tries to use arguments from science to disprove the existence of God is shown to be wrong. Obviously, if God is the creator of all natural science, and religion is the way to be connected with God, then religion is a higher knowledge than science, and religion is more important than science. Thus for these believers it is more important to understand religion than it is to understand science. If you can pray in a proper manner to God in order to influence his control of natural events, that is more important than learning the natural processes. Even if you did master the scientific processes, the creator God could change things at his whim to respond to your prayers.

The problem with this thinking is that it is mixing the games of science and religion. It is not playing by the rules of either game. The game of science does not allow contemplation or discussion of the supernatural. You cannot bring God or miracles into the discussion, and still be playing the game of science. The definition of the modern game of science is that it studies the natural processes, the regular functioning of the natural world, and it does not try to analyze or discuss the irregular and unpredictable processes of the supernatural world. This means that the Intelligent Design movement cannot talk about God when functioning as a part of science. The leading scholars in this movement understand this, and they are very clear that they are not trying to break the rules and jump to any discussion about God. But many Christian believers succumb to the temptation to make this jump. But then they are not playing by the rules, and the result is neither good science nor good religion.

The reasoning behind the Intelligent Design movement does not require the leap to the existence and functioning of a god. If you stay within the rules of the game of science, you have to approach the question strictly from a natural perspective. If the evidence we observe in nature suggests that things were designed to fit together so well, then who designed it? You are not allowed to discuss the supernatural in the game of science, so you are required to seek explanations within the natural world. You are not allowed to suggest that these things were done by ghosts, or spirits, or some other supernatural force, such as God. So what intelligence within (not outside or beyond) the universe designed our planet and our biology?

Perhaps it was some race of alien beings from another galaxy that came to our solar system a long time ago and designed the biological functioning of our world. This idea is relatively common in science fiction stories. Some science fiction stories posit a future time when the human race might be able to travel to other solar systems and "terraform" other planets. Humans would find a planet with the proper elements and the proper position relative to its star (or perhaps, with enough knowledge and power, we would even move planets into position or assemble them from asteroids), and then we would seed the planet with the proper types of bacteria and other microbial agents so that they would recreate a planet where humans could live. The microbes would interact with the rocks so that eventually they would release gases that, in time, would constitute a breathable atmosphere. Water would be released or formed, and the energy forces would create a hydrologic cycle that would make water available for living organisms. When the right materials were available, the seeds and spores of plants and other microbial life would be introduced, and they would be allowed to multiply. Along with that, at the proper time the right kinds of animals would be introduced, until the ecology of the planet matched the ecology of Earth. Then we would have another planet fit for human population. If we can imagine doing such a large task in the future, could not some alien race have done something similar in the distant past? When we think about how we might terraform a planet in the future, we usually assume we would not have to design all the microbes from scratch, but simply use the microbes we find on Earth. It is not that big a step to imagine an intelligent race that was able to design custom microbes to do exactly what they desired. Today, if you know how one computer program works, you can design another one that does something different; or if you know how to design an automobile, you can switch projects to design either a truck or a golf cart, or even a submarine. An ancient alien race might well have been able to design custom microbes.

So who is or was this ancient intelligence that designed our current world? Where are they? How can we contact them? In all of this speculation and research, we must stay within the rules of science. If the data suggested by the Intelligent Design movement suggests there is a designer, we must use the techniques of science to investigate this. Perhaps this intelligence will not have physical existence such as we have, but will be only a sort of floating intelligence, something like a super computer program that no longer needs to have a physical computer to exist, but that is self-maintaining and continues calculating using only ethereal processes such as radio waves or something more exotic. This kind of speculation moves us very close to how many people think of God. But we are still remaining in the domain of the natural. We are not talking about something that is supernatural. The end product of this kind of speculation does not give us what we think of as God in religion, but only another intelligent species. It may be smarter than us humans, but it does not necessarily have any moral authority or control over us. Furthermore, if there is one such strange kind of intelligence out there in the universe, perhaps there is more than one.

Or perhaps this intelligence is not within our universe, but in some parallel universe. Modern scientific theory speculates about the existence of parallel universes. Some scientists speculate that whenever a "choice" is made in quantum physics, an equal but opposite "choice" is also made and a different entire universe splits off and goes its own way. In that case there is a "multiverse" of many different universes, continually multiplying and creating new ones. This kind of speculation is for all purposes equivalent to fantasy, the imagination of other worlds. Parallel worlds, fairy worlds, dreamscapes, spirit worlds, inter-dimensional worlds all allow us to speculate about different kinds of worlds that might exist, and that somehow might interact with our world. Perhaps some intelligent life in a parallel world designed and created our universe as an experiment, and is observing us as we develop. But in so far as we stay within the game of science, we are insisting that this involves natural forces that we eventually hope to understand and use. Whatever beings we may imagine or discover, whether they match our thinking about fairies or spirits in some way, they are still beings, and they are not supernatural. They are not "gods" as the leading religions understand God to be today. In the television series *Star Trek: The Next Generation* there was an occasional character called Q. This character possessed intelligence and power far beyond what humans possess today, or even beyond that assumed for the space-traveling characters in the television series. But in his continuing interaction with the human characters his moral framework was shown to be less than divine, no matter how close to God his physical powers might be.

The other part of the problem with Intelligent Design is that the movement does not make for good religion. It suggests that the primary reason for believing in the existence of God is based on physical evidence. You are asked to believe in God, not because you have found some evidence for the supernatural, but because of a conclusion drawn from the study of the natural. This evidence from the natural is not some type of miracle, but merely a chain of reasoning about the natural. This comes very close to ancient pre-scientific pagan reasoning. If there is an unexplained celestial phenomenon such as an eclipse of the sun, there must be a god: listen to what the shaman tells you! If the volcano rumbles, smokes, and explodes, there must be an earth god inside: give heed to the medicine man and offer a sacrifice to appease the earth god! One must be careful here, because the sacred texts of Judaism, Christianity, and Islam all make the point that the complexity and the beauty of the natural world should point the human intelligence toward the existence of the creator or designer. But those scriptures also state that there is a limited understanding of that God that can be determined from the observation of the natural. They insist that in order to really understand and know God, you must receive revelation from the supernatural world. You cannot really know God unless he chooses to reveal himself to you.

The Intelligent Design movement is a valid part of philosophical speculation, and it deserves its role in academic discussion. But it is one of those places where the human search for truth begins to cross the line from one game to another. Suppose two people, Mr. A and Mr. B, are playing a game of chess, and a third person, Mr. X, looks at the board and then makes some noise, such as a chuckle or a gasp, and looks at Mr. A and smiles, then walks away. Mr. B, the other player, having noticed this, takes a fresh look at the board and re-analyzes his entire position because he presumes this third person, Mr. X, saw something about the strategy that he has not seen. After some time, he smiles, makes a move, and the game plays to an end with him winning. Then the first player Mr. A says, "If Mr. X had not given you a hint, you would never have seen my plan and I would have won." At that point we have moved beyond the simple rules for chess strategy to the larger framework of outside interference. We have moved from the specific rules of chess to the rules of gamesmanship. In chess, no one is allowed to give comment to the players or give any encouragement. In contrast, in athletic sports it is common for the spectators to be expected to give advice and encouragement to the players; that's why we have cheerleaders. When we raise questions from the standpoint of the Intelligent Design movement, we are doing something that plays within the rules of the game of science, but which also makes a move beyond science into the game of philosophy, or the game of human reasoning. One of the questions that is raised by the Intelligent Design movement is whether we would ever be able to learn or know anything about the designer, just from studying the design. The example of a GPS system was mentioned above as evidence of something that was designed and not a product of accidents. But from analyzing a GPS system, how much could you learn about the human beings who designed it? Do the designers have wings or gills? Are they bipeds or quadrupeds (two-legged or four-legged creatures)? Do they have hair or feathers or scales, or none of these covering features? Even if we are convinced of the existence of a designer by the evidence put forward by the advocates of the Intelligent Design movement, that does not necessarily imply that we can learn much about the designer. Staying within the rules of the game of science means we must conduct that search within the framework of the game of science, and not shift to the game of religion. But it does not guarantee that we can find answers. As mentioned, scientific speculations about parallel universes converge with fantasy speculations about fairy worlds and alternate universes. When we move beyond the framework in which we can observe, demonstrate, and perform experiments, we move beyond the game of science. There is more to the wonder of human existence than just the game of science as we have defined it and play it today. The questions that come up are fair questions for humans to discuss. But those additional questions move us outside the boundaries of the structured game of science. In the example of the chess game offered above, the first player asserted that unless the third person had interfered, the second player would never have spotted the strategy and won the game. How can we know that is true? It is possible the second player would have spotted the strategy anyway. But we have no way of investigating that or proving it. We can only move into speculation about what we consider the potential of another person's mental habits. Then it is not clear if we are talking about chess or about psychology. So the Intelligent Design movement has a valid place within the game of science, but it also raises questions that move us outside the game of science, into an area where other data coming from both religion and philosophy participate in the discussion. But at that point it is outside the simple game of science, as well as outside the simple game of religion. Within the game of science, Intelligent Design cannot be used as an indication of the existence of God or of the supernatural. Looking at it from the perspective of the game of religion, Intelligent Design does not offer any more indication of the power and existence of God than what is indicated to the believer by the basic facts of biology, physics, or cosmology. Within the game of philosophy the Intelligent Design movement offers some very interesting ideas to consider. These include the question of what constitutes a design, what is the significance of finding design in the structure of the universe, what is the significance of the existence of the universe, what is the significance of the existence of human beings in the universe, what is the significance of one human in relationship to all other humans, and what can a physical creature perceive and understand about the universe, or what can a physical creature perceive and understand about God? But the game of philosophy is not governed by the rules of science, nor by the rules of religion, and it takes us beyond either of these games that we have devised to try to determine truth, that are the subject of this book.

Chapter Eight: Design and Evolution

In the previous chapter I dealt with the problem concerning the Intelligent Design movement. In this chapter I seek to balance that criticism with a positive appreciation for the issue of design. But I also intend to demonstrate the positive role of the discussion of evolution within the game of science, and to build a bridge between design and evolution that is not contradictory to the game of religion.

For as long as we have any records, human discussion of the existence of a god has included, as part of its reasoning, the matter of the apparent design in the natural world. The functioning of the forces of nature has seemed to fit together so well and in so complex a manner that it does not appear to us that the world could have come about by accident. We do believe that some things result just by chance. For example, the forces of erosion have created dramatic, beautiful sculptures in some parts of the world, including arches, pillars of rocks, and strange canyons. But there is a difference between those natural features and those created by human beings. In the wilderness of the American West, a common item used by humans to mark significant points on trails is a stone cairn. This is a simple pile of rocks. Despite the fact that there are many dramatic rock formations created by the chance forces of erosion, a human-made pile of rocks is generally very evident. The height and shape of the rock pile made as a cairn is different from anything in nature. Or at least it is in theory: in practice, a human-made cairn sometimes can be very simple, and that can be confusing. Is a pile consisting of two rocks a human-made cairn, or a natural feature? On some of my trips with Boy Scouts we devised a simple rule: it had to be a pile at least three rocks high to be considered as a cairn. Even then we sometimes had to study things closely. Were there other similar piles nearby? Was the material and shape of the rocks of such different character that it could not be the result of slow erosion of layered rock? But we never got lost because the vast majority of the rock piles were easy to sort as to whether they were natural or human creations. In a similar way, we can look at exposed layers of sedimentary rock, and we can note the beauty that often results from the sculpting due to water and wind erosion. But we have little trouble sorting such natural features from the places where human machinery was used to cut through the hills to allow a road to be built. We can detect when things are shaped in such a way as to have a purpose convenient to humans, and are not simply the result of random natural forces. So when humans look at the complex regularity of motion found in astronomy, or the many aspects of plant and animal biology that follow the regular cycle of the seasons, or the beauty of sunrise and sunset, in which the aspect of beauty as far as we know is purposeless for plants and animals, we begin to suspect that this complex coordination hints at a designer with some greater purpose in mind.

The argument of apparent design indicating the existence of a god changes over the years as human knowledge grows. Let me discuss cosmology first. As our knowledge of astronomy grew, we came to understand that the solar system and even the galaxy basically consist of floating rocks, that is, rocks and dust floating in the force fields of gravity. But the vast scale of the universe raises questions about its purpose and function. At first when we learned about the huge scale of the solar system, the galaxy, and the existence of other galaxies, it made our planet seem so insignificant. But the more we learn about the forces of physics at work in our universe, the more we appreciate how everything has to exist at such vast distances in order for their function to work properly in regard to humans on Earth. The stars have to be very far away so that they remain in effect stationary in the sky, and can serve to mark time and to assist in navigation. They have to be huge to shine from so far away. They have to be huge to burn for so long, so that humanity can experience a period of apparent stability in its observation of nature. They have to be far away so that they do not cause disturbing gravity effects on Earth. The moon must also be far away, and yet close enough to cause the tides. The tides are necessary to help cycle energy received from the sun to other parts of the Earth. The moving super-heated metal inside the Earth is necessary to create the electromagnetic fields that shield the Earth from the solar blasts of radiation and particles that would otherwise sterilize the planet. On the one hand this is clearly all just random chance of floating rocks and burning stars. But on the other hand, why does it fit together so perfectly to form the world on which we live? And why have we not found anywhere else where such a planetary condition exists? From our current observations, even if we eventually find other planets similar to Earth, it appears right now that this might be a rare situation.

The advances in the sciences of biology and ecology also continue to hold out this taunting suggestion of design. Prior to our knowledge of cellular and microbial life, humans were impressed by how animals and plants lived together in dependable relationships in nature. When we learned about microbial life (which required microscopes, which required glass lenses, which required the invention of glass after the development of primitive metallurgy), we came to understand so much more about the human body, about disease, about reproduction and child growth, and about what we call environmental science today. On the one hand the vast quantity of different creatures and the knowledge of how they function in nature make the human being seem so irrelevant and seem like such an accidental product of some biological evolution. But on the other hand, the appreciation of how the entire system functions, and continues to function apart from human guidance, and continues to function smoothly in many cases in spite of human interference, begins to suggest the parallel of a carefully designed nursery room where a child has space to play and grow, but is protected from accidental extinction or from destroying the nursery room.

In one sense, the argument from design (that the universe was created by God and not by chance) continues to keep pace with the advances in scientific discovery. When our knowledge of natural things was limited, we were impressed with the complexity and intricacy of certain things, and they suggested the possibility of significant planning behind their arranged system. When we learned more, we sometimes came to understand some details that were once a mystery. But with this greater knowledge came awareness of greater mysteries. Take as an example our research into the smallest of things, the opposite of studying astronomy. At first we could only look at tiny rocks and tiny insects. The ancient Greek philosophers speculated about atoms, which they identified as the smallest units of matter. The ancient Buddhist scholars speculated that if you kept splitting pieces of matter, you eventually got below the size of the smallest bit, and so they theorized that in reality all matter is made up of nothing. More modern scientists eventually worked out the theory and structure of molecules, and deduced that molecules were made of atoms. But then they went to something smaller than atoms when they figured out that atoms themselves consist of collections of protons, electrons, and neutrons. Then they built machines to split these pieces, and began to theorize about quarks and bosons. In the meantime they also learned that atoms can have isotopes: similar atoms with different counts of internal particles (mainly neutrons) that change some aspects of the way the material behaves. The recent identification of the Higgs boson by scientists at CERN seems to support certain modern theories about the different characteristics of the different bosons, with the Higgs boson responsible for the quality we call mass. The other bosons are considered particles, but they do not have the attribute of mass. If the other bosons do not have what we call mass, were the ancient Buddhist scholars correct in their speculation that all matter is really made up of nothing? The more we learn, the more questions we have, and the more marvelous the entire system seems.

Let us take another example of research into small things. In the area of biology and chemistry, we humans moved from studying tiny insects to studying microbial life, and cells. Then we began to look at the interior pieces of cells. Scientists built theories of the master control molecule, and eventually we discovered DNA—deoxyribonucleic acid. Once we found that, we thought we would have the key to all the building of cellular structure. We determined that the parts of DNA were combined in sections to make up genes, which controlled certain aspects of the design of cells. But today scientists have developed the study of epigenetics, in which they believe some parts of the DNA strand also control which other parts are turned on or off ("expressed" in the functioning of cell construction). Some of these control parts can be affected by environmental circumstances. There is a long-standing argument over which factor is more important in the final shaping of a living being: nature or nurture? Nature refers to the genetic content which makes up a person, and nurture refers to the formative influences from the outside world as the person grows. This argument does not end; it continues to be argued. A human being is the result of the design of his genes; but the genes may be affected by the environment, or by the environment of his parents. In science we find that it is true that the more you know, the more you realize you do not know.

The more you learn, the more questions you raise, and the more mysteries there are. In cosmology we went from conceiving of a single planet with a sun circling around it to a conception of a solar system with this wonderful planet orbiting the sun. Then the stars were identified as additional suns, very far away, and we conceived of this huge assembly of stars called a galaxy. But in the twentieth and twenty-first centuries we realized that there were many thousands of additional galaxies, and that even galaxies might exist in clusters of special shapes. In atomic physics we went from molecules to atoms to electrons to quarks and bosons. In biology and chemistry we went from animals to cells to interior cellular structures and to genes, only to come back to the study of what we call stem cells that can be activated to turn into almost any other kind of cell within a creature. All of this information increases our awareness of the mechanical functioning of the details of our universe. But it also increases our appreciation of the complexity of this interaction of matter and energy, and that appreciation of the complexity continues to raise the suggestion of an intentional design by some creator. The history of this argument from design suggests that no matter how much further our scientific learning takes us, the question of design will keep pace with the learning.

The Intelligent Design movement discussed in the previous chapter is, in that case, merely a new and current expression of the argument from design. In the debate about religion and the existence of a god, the Intelligent Design movement offers a new focus for the evidence of design against the argument for mere accidental mutation. If some of the older arguments for evidence of design seem to have been refuted by more recent research (arguments such as the distinction of species, or the design of the eye), the state of modern research offers other pieces of evidence that seem to point to what Intelligent Design proponents call "irreducible complexity."[15] Perhaps in the future new information will allow some scholar to offer a theory that seems to explain the new evidence, offering a chain of "chance" mutations that resulted in the current "irreducible complexity." But if so, it is highly likely that such new information will also reveal new mysteries that will themselves point to what appear to be more hints of intentional design. In the area of religion proponents of both sides of the argument about the suggestion of a designer will feel they are winning: those who argue that the evidence from design has been refuted, and those who argue that the evidence from design has become more complex and stronger.

But there is another aspect to the phenomenon of design that must be discussed, and this is connected with the theory of biological evolution. The matter of design is *essential* to the theory of biological evolution. This is something that is often not appreciated by advocates of religion. In this sense, it can be said that the theory of biological evolution can be used as an argument for the existence of God, rather than against it. Or to be more precise, the connection between design and the theory of biological evolution places us exactly at the position just discussed in the preceding paragraph: the matter of design will be evaluated by the individual person in accordance with his other beliefs, therefore as supporting either the argument against a god or the argument for a creator god.

---

15 The reader can compare chapter 3 by Mark Ridley, "The Mechanism of Evolution" and chapter 24 by Michael Behe, "Evidence for Intelligent Design from Biochemistry" in James B. Miller, editor (2001), *An Evolving Dialogue: Theological and Scientific Perspectives on Evolution*, Harrisburg, PA: Trinity Press International.

In order to discuss this relationship between biological evolution and design, we must distinguish between two uses of the term, which are sometimes called micro-evolution and macro-evolution, or evolution on the micro scale, and evolution on the macro scale. The term micro-evolution refers to the phenomenon that small characteristics change in living beings from generation to generation. There is no debate about this phenomenon. Everyone agrees that it is a fact that in the reproduction of living things, small changes can and do occur. In bisexual reproduction, this is part of the purpose: the characteristics of both parents are contributed to the potential characteristics of the child, and a selection happens by some natural process, so that the child that results is a combination of the genes of both parents. The child is not exactly like either parent. In human beings, the child may inherit his father's height and his mother's hair. This phenomenon of micro-evolution underlies the entire science of breeding crops and animals. The genes of parent plants or animals are combined to produce descendants with a new set of characteristics, and then, in the case of plants and animals, the desired combinations are used for the breeding of more generations, and the undesired combinations are discarded. In animals or plants that reproduce by asexual reproduction, you do not get this change of genes and thus a change of characteristics. But there are other phenomena that cause changes in genes. Among these are radiation, and chemical accidents. Some types of radiation can affect the structure of the DNA molecule, resulting in a change in the chemistry of this molecule. Then when the creature reproduces, the new offspring will manifest this change as it develops. Chemical accidents can happen before, during, or perhaps even after the reproduction process. Somehow something goes wrong in the building of more strands of DNA, and the result is a molecule with something different from the parent DNA. The resultant offspring again develop characteristics different from the parent. These types of changes in the chemical structure are called mutations, and they are very important in the development of some microbial life. In particular today we worry much about how certain bacteria and viruses mutate over time so that they accidentally develop immunity to chemical agents that formerly killed such microbes. Thus certain bacteria became immune to penicillin, and we are unable to cure a person infected with the disease caused by that germ. The phenomena that cause mutations in creatures using asexual reproduction can also affect creatures using bisexual reproduction. As stated above, this phenomenon of micro-evolution, or tiny changes in the genetic structure, is well known and acknowledged by everyone. Both in sexual reproduction and asexual reproduction, tiny changes happen.

    The topic of macro-evolution is a different matter, and this is the aspect of biological evolution that is hotly debated, particularly by advocates of certain religious positions. The theory of macro-evolution says that eventually enough mutations may accumulate in the DNA of a creature so that it becomes a different species, separate from the original species. While some descendants of the original creature may continue to reproduce and exist in an environment while maintaining the original characteristics, other descendants slowly come into existence through the reproduction of generations that contain more and more mutations, and thus possess somewhat different characteristics. At some point the difference in the DNA structure between the two sets of descendants is such that they cannot mate and reproduce with each other. Instead of one species, we now have two. The grand scheme of the theory of macro-evolution says that somehow, beginning with one original strand of DNA, this process of mutation leading to a new species has occurred over and over again countless times, slowly leading to the wide distribution of the wide variety of plants and animals that we have today, as well as many others that have become extinct over time.

The first half of the theory of macro-evolution, the part about how a new species might develop from mutations, is not really a subject of debate. Everyone agrees that this process can happen, at least potentially. The second half is where things become more complicated. The second half offers speculation as to how often and under what circumstances this production of new species takes place, and it theorizes backward from the current variety of species to one original ancestor. There is strong disagreement over the idea that there was one original ancestor, and that all living creatures are mutations that evolved from a common ancestry.

Before we knew about microbial life, genetics, DNA, and chemical mutations, we could only speculate about how life forms might be related. It was obvious to anyone that there was a fairly common pattern among many life forms, and that we were made of the same materials, or today we would say the same chemicals. Animals ate plants and other animals, and humans ate plants and animals. We were made of the same materials, and the materials or chemicals of other living things were food for us. Many creatures followed the common structure of one head, four limbs, two eyes, two ears, a skeletal structure with a backbone, skull, and hips, etc. There was obviously a common pattern to the design of many species. All mammals had some hair, and nursed their young with milk. Other creatures laid eggs and hatched them; some fed their young, and some did not. Some creatures were very different from this common pattern. Snakes had no limbs. Fish were sort of similar, having backbones, but lacked limbs, though the fins were sort of similar. Birds had feathers instead of hair, and wings for their forelimbs. Insects were quite different, with six limbs (and arachnids such as spiders had eight legs; and then there were centipedes), and creatures such as starfish and jellyfish were even more different in limbs. But the similarities between humans and animals such as bears, monkeys, wolves, dogs, and cats were enough for humans to speculate about their connections, and to wonder in what sense the creator god or creator spirit had used a similar pattern. As we developed the science of anatomy, this curiosity became greater. It turned out that snakes had four vestigial limbs, as did dolphins. Many fish shared the characteristic of the skull and backbone, but some did not have a backbone. Knowledge of the changes in form undergone by certain creatures, such as tadpoles to frogs, or caterpillars to butterflies, allowed speculation about how all creatures must develop from microscopic size to their eventual full size. Then global exploration added to the pool of information. As the world was explored by the Europeans, more and more species were discovered that were very similar one to another, but having small differences, and it was noted that these different species often lived in different locations from each other. There was a great effort by naturalists to catalog all of this data.

The problem for scholars was how to handle all of this data about so many living creatures. How could you group them and classify them so that important similarities were identified, and how could you explain why there were so many varieties of similar creatures, sometimes with nearly identical bodies and yet strikingly different characteristics. For example, why are some snakes venomous, and others not? The important key to resolving all of this complexity is generally attributed to the theory of Charles Darwin in the nineteenth century, though it should be mentioned that many other scholars were involved in the discussion and some had proposed similar ideas. Darwin was fortunate to participate in a trip around the world with the express purpose of studying the different types of flora and fauna he encountered in as many locations as possible. This opportunity allowed him to collect a large amount of data, and the time to study it and think carefully about what he was observing. As he reflected on the different varieties of plant and animal life that he collected, and as he thought about how each functioned within its particular environment, he developed a theory called the **theory of natural selection**.

Today, many people think that Darwin invented the idea of the evolution of different species. He did not. Darwin and others simply accepted the idea that somehow changes occurred over time in the forms of plants and animals. Even though the science of DNA molecules as the carrier of genetic information had not yet begun, most scholars or naturalists accepted the idea of species differentiation and evolution, at least to some extent. They just did not know how these changes would happen. The well-known facts of the breeding of animals and plants made it obvious that very similar species must somehow have some common history. The question was how to explain all of this variety, and at the same time explain the similarity. Why did you find some varying species of some creatures, and not find all kinds of every possible variation? For example, you find black bears and brown bears and white bears, but no green or pink bears. Random mutation and variation should create every possible variation of every kind of creature, in every location. But instead you only found certain species with certain characteristics, and these different species were often located in separate geographical areas (brown bears in the forest, white polar bears on the pack ice of the arctic region). Darwin produced a theory to explain, not how the changes occurred, but how the different characteristics resulting from the changes *interacted with their environments* so as to influence which types of changed creatures survived and reproduced, and which ones did not survive.

When humans breed animals, they look for certain characteristics that they want to continue, and they discard the offspring that do not have those desired characteristics. Darwin produced a theory that explained how the same sort of selection happens in nature, as a natural process, without having to have a purposeful, intelligent breeder making the decisions. This was the theory of natural selection. The following discussion will use the example of animals, but the same concept applies to the variation of plant life. When mutations occurred, the resulting characteristics of the creature produced a certain interaction with the creature's environment. Some changes increased the creature's ability to prosper in its environment. Perhaps it was a bit faster, or taller, or had a longer reach to obtain food. Perhaps it was more sensitive to finding prey or avoiding predators. Whatever the characteristic, if it helped the creature prosper, then it also led to that creature being more successful in reproduction. If it did not help the creature prosper, then that characteristic might contribute to those variants being eliminated by becoming prey to other animals, thus reducing the frequency of reproduction for that creature. It was the fit between the creature's environment and the creature's characteristics that made the difference. The creatures with the better fit got more food, avoided predators, and reproduced their characteristics. The creatures with the less-advantageous fit did not prosper, were caught and eaten, and did not reproduce. In short, nature itself served as a breeder, selecting for the characteristics that fit the environment. This is the theory of natural selection, and this is the great insight of Charles Darwin.

The theory of natural selection provided a logical system for understanding two things. It explained the relationship of the creature to its environment. It also explained the process of the "natural" breeding of creature characteristics, explaining why some variations survived and thrived, and others disappeared. With these two pieces of logic, scientists finally had a way of understanding why the different types of animals existed, and why they were distributed in different parts of the world. Note that nothing in the theory of natural selection explains how the mutations or variations occurred. That was simply assumed as some natural part of the reproduction process. Darwin's theory of natural selection is not about the process of the change or evolution of the characteristics of the animal. It is about the process of the survival of the animal within its environment.

Darwin finally published his book *On the Origin of Species* in 1859, and from that time on the study of biology had to require the study of how a creature fits into its environment. The characteristics of the animal could only be fully understood by taking note of how those characteristics made it fit to function well and prosper in its environment. But if we are talking about how a creature is so constituted as to fit and function within an environment, we are talking about design. The process of natural selection provided a mechanism by which animals were constantly improved in their design to better fit into their particular environments. As an environment changed, there was a corresponding need for a change in design, and the process of natural selection enabled that change in design. An environment might change for many reasons. Population growth might make competition for food more difficult; wandering in search of food or more room might lead animals to migrate to other islands or forests or prairies or parts of the sea; new predators might enter the area and change the situation with regard to safety; weather and climate changes might create new conditions for selecting dens or nesting sites, or change the type of vegetation or food sources; the growth, or death, of forests might change the entire vegetation system of the area. Whatever the reason, the process of natural selection enabled a constant breeding program that allowed the design of the creature to continually change to fit the new environment.

The theory of macro-evolution demands a role for design within the animal and plant kingdoms. The modern science of biology is completely based on this study of design. The theory of natural selection provides a rationale for how each design is tested and for how the competition winners are selected. If you do not allow this theory of macro-evolution to be discussed, the modern science of biology cannot function. It has no other framework for how to study plants and animals. The theory of macro-evolution provides a framework for analyzing how a creature fits into a particular environment, and for analyzing how each part of a creature functions with regard to life in that environment. In doing this it also provides a framework for analyzing change within the creature, within the species, or within the collection of related species, by how they are modified with respect to the environment. The theory of macro-evolution makes modern biology possible by providing an explanation for the process of design in the plant and animal world. If you do not talk about the design relationship between characteristic and environment, you have no framework to talk about biology. You might as well write a dissertation on why dolphins fly to the moon. A plant or animal only exists, and exists in a certain location, because it was designed to fit in that location. A biologist studies this design, and explains it to other students.

For a religious biologist, this study of the design in biology is nothing less than the study of God's design. For a non-religious biologist, this is the study of the design found within nature. The non-religious biologist does not believe that the identification of design in nature points to the existence of a supernatural god. For him, the design "just is." The topic of the supernatural is not part of his conversation or his thinking. He may well speak in admiring terms of the beauty and the wonder of the design found in the universe. But he is not linking the creator because he does not label the universe as the creation of a designer.

This is not a matter of breaking the rules, of mixing the games of science and religion. Within the game of science, the supernatural is not considered, and not discussed. That is why Darwin's theory of natural selection is so important. Without it, you have no way to describe and analyze the structure that exists in biology. Without it, you cannot talk about any design without crossing into a discussion of creation and the supernatural; that is, you must switch into the game of religion, and that is against the rules of the game of science. But with the theory of natural selection, you can remain within the rules of the game of science, and talk extensively about how well designed the world of biology is. When you move to a discussion of the Designer, you shift out of the game of science. Just as discussed above in connection with Intelligent Design, when you move to consideration of the Designer, and you choose to identify the Designer as God, you change games. It is perfectly okay to change games and play a different game. But you must inform your fellow players that you are changing games, and you must allow them to refuse to play the other game with you.

In the next paragraphs I am going to bring information from both games, science and religion, into play to discuss how it is not necessary to assert a conflict. I will not mix the games here. Rather I will try to illustrate how neither game dominates the other.

The two parts of the theory of macro-evolution that upset some religious people are, one, the backward theorizing—the tracing of all life backwards in its evolution to the one original life form—and two, the linkage of the creation of human life as a simple extension of this same process. Note that neither one of these parts are essential to the main theory. You can speculate everything descends from one original strand of DNA, or that somehow life on Earth began with several different strands of DNA, somehow formed at the same time, and each of them then developed through history into its own particular kinds of plants and animals. And you can speculate that God used the regular process of mutation and natural selection to bring about humans, or God performed a special miracle of additional creation.

The two reasons to assume one common original ancestor for all hominids and humans are the similarity of DNA, and the simplicity of the argument of one source. On the other hand, however one envisions the original formation of the first DNA, it seems possible that you could also envision several variants of that complex chemical coming into existence simultaneously.

As for the human race, the question becomes whether God controlled "natural selection" to evolve the human line, or whether God intervened by an additional miracle to suddenly create a new human creature from mere molecules. Everyone is at liberty to theorize about the relation of modern humans to modern apes and to other hominid fossils. The Old Testament, the sacred scriptures of Jews and Christians, states that God created both animals and humans. God did this at essentially the same time (on the same day, according to Genesis 1, and also apparently in Genesis 2), and in a similar manner (both made of earth). These scriptures say that God did some additional distinct act when he created the human. So you must account for the similarities and the differences between humans and hominid animals. Current paleontological scholarship says that modern humans, *homo sapiens*, interbred with both Neanderthals and with Denisovans (a new hominid identified from bone fragments found in east Asia).[16] Accepting that scholarship, each position in this discussion must decide at what point God made this crucial distinction between animals and humans, and what that distinction was. As far as the science is concerned, it makes no difference whether God supervised a crucial genetic evolution out of the ape line of hominids, or whether God took some dirt and designed the new creature from scratch. If God created humans from scratch, it would appear that when God built the new human creature, biologically the new creature was so similar to the previous ape and hominid creatures that we today cannot easily point to any biological difference of significance. Whether God supervised a genetic evolutionary change, or built a nearly identical creature from scratch, at some point in Earth's history, at God's choosing, the Biblical human being appears on the scene, and that leads to human history.

Since these two parts of the grand theory of macro-evolution (a single DNA source, and the evolution of humans) can be envisioned either way in the theory, it is perfectly possible to

---

16   This new discovery is discussed in Jamie Shreeve, "The Case of the Missing Ancestor," *National Geographic*, vol. 224:1 (July 2013), 90-101. However, a new theory is being proposed at the time of the writing of this book that suggests that many hominid fossils might not represent different species, but simple variation of characteristics within a species. Thus the assumed evolutionary chain of hominids leading to modern humans might consist of a much smaller group of creatures than the previous theory proposed. See "Pruning the Human Family Tree" in the *Austin-American Statesman*, October 18, 2013, byline John Noble Wilford from the *New York Times*, reporting on an article by David Lordkipanidze, Marcia S. Ponce de León, Ann Margvelashvili, Yoel Rak, G. Philip Rightmire, Abesalom Vekua, and Christoph P. E. Zollikofer, "A Complete Skull from Dmanisi, Georgia, and the Evolutionary Biology of Early *Homo*," *Science*, 18 October 2013, vol. 342, no. 6156, pp. 326-331, DOI: 10.1126/science.1238484.

hold either position on either question and continue to function as a good scientist in the world of modern science. You can argue for a single ancestral creature, or for several original strands of DNA. You can argue for modern humans being created by God as a continuation of the evolutionary chain of hominids, or you can argue for a special miracle of separate creation of the human that used the same creature pattern God had already designed for animals. Neither of these variations of the theory of macro-evolution contradict or cancel the theory itself, and it is unlikely that some kind of scientific evidence will ever be found that would prove either position of each question. The reasons for holding to the variants of multiple DNA strands at creation and for a special act of human creation do not come from the evidence of science, but from the information interpreted from the sacred scriptures of the Jewish and Christian religions. This comes close to violating the rules by mixing the play of the two different games of science and religion. But it does not quite cross the line. The information from supernatural revelation is used to speculate on some detail of the scientific process that occurred long ago and that is most probably beyond testing. The scientific reasoning about the ancient process prefers to assume a simpler happenstance, but in so doing it runs the risk of contradicting the revelation from the supernatural world, in the opinion of some interpreters. As long as all players keep straight in their minds which parts of truth they are obtaining from each separate game, and they do not let their reasoning break the rules of either game, everyone can continue playing each game together with their colleagues.

If the theory of macro-evolution is all about design, and if you believe that the evidence of design in nature points at least in some minimal way to the existence of a god, then the theory of evolution is evidence for religion, not against religion. You cannot mix or confuse the games of science and religion. But within the game of science, all of the talk about evolution does not reject or contradict any of the value of the game of religion. If we go back to our example of the game of checkers, the theory of evolution stays within the black squares. It does not move onto the red squares, nor does it say they do not exist. The theory of evolution gives a framework for modern biology to study the logic of living things. That logic allows the discussion and appreciation of design in biology. The recent Intelligent Design movement is then simply one more outgrowth of the theory of biological evolution, and not contrary to it. And the work of modern biologists has nothing intrinsically anti-religious associated with it. Indeed, the work of modern biologists would seem to be consistent with those sacred scriptures that suggest the recognition of design in the natural world should be recognized as evidence of the handiwork of the Creator.

Darwin's theory of natural selection is essential to the modern science of biology, and it is not in contrast to religious scriptures. The debate is not about how biology functions, but about when and how God set the initial conditions that have evolved to our modern world. Did God the Creator begin with a multitude of different creatures, which all evolved according to their breeding lines, or did God begin very long ago with a single piece of DNA and take his time using natural selection to bring about his design for all the current creatures on Earth? For religious people, both options are possible, and both options require the assumption that God the Designer planned into his creation the entire process of natural selection and all the breeding choices that natural selection has made.

In the Christian New Testament Jesus states that "not one [sparrow] will fall to the ground apart from the will of your Father" (Matthew 10:29). This has enormous significance for a Christian understanding of the process of biological evolution. Let me use an analogy regarding politics. The Christian scriptures teach that the rise and fall of empires was all planned and controlled by God, and continues to be so controlled (see passages in the books of Daniel, both halves of the book of Isaiah, the foreign nation oracles in the books of Jeremiah and Ezekiel, and consider other passages in the books of Genesis, Exodus, or the Psalms). But the rise and fall of temporary human governments is not the major focus of God's plan, which is his creation of his church of disciples. The human governments are just tools used by God to help run his creation (see Paul's comments about government in the book of Romans). In the same way the rise and fall of species must be understood as part of God's plan, completely under his control, all designed to fit into his larger plan to serve his church of disciples. God controls throughout all of history which creatures live, and which die. God controls the process of natural selection. Natural selection becomes God's tool for creating the world of living things.

But note that placing all of this under God's control does not remove the possible negative role of human sinfulness. Humans can and have sinned in managing human government, bringing about God's anger and punishment, and humans can and have sinned in managing the care of God's earthly environment, resulting in God's anger and punishment, as attested often in the words of the Israelite prophets (see the books of Joel, Hosea, and Amos to start; see also the significance of nature in Deuteronomy 27-28). Indeed, the original commission from the Old Testament God to humans was to tend and care for God's garden, and the beginning of sin involved the misuse of certain fruit of the garden. Therefore the Christian religion makes humans responsible for assisting in the management of the natural world in accordance with God's desires in the same way humans are to assist in the management of human affairs in accordance with God's desires. God always has ultimate control, but humans are given a subordinate control, making them God-like in their management over things (see Genesis 1, or Psalm 8, or Psalm 82).

The understanding of God's use of biological evolution (both micro- and macro-evolution) will depend on how the believer interprets many parts of his sacred scripture, and this chapter of this book is not about the detailed discussion of that interpretation. This chapter is about the positive value for the game of religion that comes from the argument of design as observed in the natural world. That evidence of design is a crucial part of the modern science of biology. The process of natural selection in biology is a process of the continuation of design, with the new design always adjusting to fit the changes in environment. Whether one understands that God used this process in only a minor way, to make small adjustments in plants and animals in recent millennia, or whether one understands that God used this process in a larger way as a method of taking some chemicals from the Earth and letting them gradually form into the complex current world of plants and animals over billions of years, the observation of the value and the complexity of this process of design in biology is impressive for a religious person. For the religious believer, the process of natural selection in biology becomes part of the grand design of the universe. At that point the process of natural selection creating properly designed variety among plants and animals becomes parallel to the process of the creation of humans from the combination of sperm and egg cells. The believer understands that new humans don't simply pop into existence, but they are grown in a complex process inside the wombs of their mothers. Just as a seed is not a full tree or even a flower, so a fertilized human ovum is not a full-grown human being. Yet somehow that original human cell in the womb grows into a normal human being with all the appropriate organs, with a sensing and thinking brain, and with his own personality. Even after birth the human child still has to develop until it becomes a mature man or woman, with the distinct characteristics of its gender. Believers marvel that God designed this process of growth and maturation that happens automatically. Religious believers who study biology marvel that God designed a similar process that functions automatically to continually develop plants and animals to fit into constantly changing environments. The increased knowledge of the process of human reproduction has not diminished the sense of awe and admiration believers have for God's use of the birth process. The increased knowledge of the function of the process of natural selection in evolutionary biology has not diminished the sense of awe and admiration believers have for God's use of the process of biological evolution.[17]

There is one last point to discuss before concluding this chapter. The idea that the process of natural selection allows the process of biological evolution to function automatically is perhaps the most controversial aspect of the tension between science and religion in connection with biology. For the non-religious person, this process of natural selection would seem to negate any need for any miraculous manipulation or intervention by any god in the process of the evolution of plants and animals. For over a century this idea, that a god is no longer necessary to understand the functioning of the complexity of nature, has been resisted fiercely by some religious believers. The argument of this chapter is that this debate is no more necessary than the older debate about whether the sun circles the Earth, or the Earth circles the sun. At one point many people felt that to deny a

---

[17] Research into and development of the theory of fractals and fractal patterns (the natural recurrence of similar patterns in physical phenomena) has certainly also contributed to an enhanced appreciation of the possibilities of how God might design a system that automatically develops itself, but still follows principles or rules designed into it initially.

geocentric view (in which the Earth is at the center and the sun circles the Earth) would be unfaithful to their sacred scriptures. The scriptures speak of the sun moving, not the Earth. Therefore those scripture passages must decide the scientific question. But that is to mix the games of science and religion. You use science to answer a scientific question, not theology. The relationship of the motion of the sun and Earth is certainly a scientific question about nature, and not a question about God. When you turn to the game of religion, the interpretation of the sacred scriptures requires a simple literal reading of the texts. You cannot turn the purpose of the sacred scriptures into a science book. We still speak today about sunrise and sunset, for we speak from our perspective as inhabitants dwelling on the surface of the planet. If this is normal speech today, then we cannot press similar speech from a few thousand years ago to have any deeper scientific meaning. Today everyone acknowledges the science concerning the structure of the solar system and the orbit of the planets around the sun. We no longer feel that we are losing something or diminishing our theology by accepting this science. We hold that we are still being faithful to the sacred scriptures. In the same way it seems appropriate today to recognize that accepting the information from modern biology such as the theory about natural selection does not produce a genuine conflict with religious faith. The claim of non-religious people that science has proven there is no such thing as the supernatural world, or no supernatural intervention in our world, has no foundation. Science cannot speak about the supernatural, either positively or negatively. If we have discovered more of the process of how biology works, and that it works "automatically," this does not in any way negate the concept of design, or the role of a creator, or the interface with the supernatural. To use another example, the fact that we know much more about weather and about light refraction does not make the sunset any less magnificent in its beauty. Certainly, we know more about how this beautiful sunset is created in the sky. But that does not deny the existence of a creator god. It only makes us appreciate more what he designed into the process of beautiful sunsets. The fact that we can now appreciate and understand more about how a creator god designed biology to work throughout the course of history does not constitute data that would argue for the negation of the creator's existence. The science does not speak one way or the other about the existence of a supernatural god. A religious person such as a Christian is not unfaithful to the sacred scriptures when he asserts that his God created plants and animals, and that his Christian Bible gives only a simple, non-scientific statement asserting this creation, while modern science provides more insight into how the Christian God designed biology to work. The argument of apparent design in nature suggesting a designer still stands. The evaluation of this argument can go only so far in the game of science. The game of religion picks up this argument, and religion provides a source of supernatural information that shapes a more full understanding of this argument of design. But as long as you want to play with other people, you must clearly distinguish which game you are playing at any given time, and stay within the rules of that game.

Chapter Nine: Prayer, Probability, and the Problem with Irresponsible Prayer

Prayer is one of the situations in which the games of science and religion frequently come into contact, so this deserves some attention in the discussion of their relationship. In the game of religion, prayer is when humans talk to God. The conversation can include many aspects, such as thanksgiving, praise, confession of guilt, or just the heart-to-heart sharing of friendship. But the name "prayer" comes from the fact that often part of the communication includes a request, a petition asking for some help from the divine world.

It is quite proper to request help from God, in Christianity, Islam, and Judaism, and in most religions. Because he is God and because he is supernatural, God has the ability to intervene within the natural world in a variety of ways that can assist his disciples. God wants to help his disciples, and God has asked his disciples to request his help. In this book I am primarily discussing those religions where the supernatural God has revealed himself to his disciples, and that revelation itself constitutes an act and declaration of grace. We can know nothing about God or the supernatural world unless the supernatural reveals itself to us. But the game of religion is founded on the fact that God has spoken, and spoken for our benefit. God has reached into our world and offered his help. In most religions it actually dishonors God if the disciple does not pray and seek God's help. As one Christian hymn puts it, "Thou art coming to a king; large petitions with thee bring, for his grace and pow'r are such, none can ever ask too much."[18] Prayer is a good thing to do.

God can respond to prayers for help in many ways. He can simply enforce or reinforce the sequence of the natural events and the natural powers that he has designed, which he does quite often. The prayer "God, help me to sleep well tonight, to be well-rested and refreshed tomorrow, and to do a good job in the presentation I have prepared to make before my coworkers tomorrow" involves such reinforcement. The person has done all that is proper for him to do, he has done his work and prepared well, and he is also tending to his personal health and well-being. He simply asks that God would superintend all this preparation so that it works normally, and that God prevent anything extraordinary from interrupting the process of preparation and the delivery of the presentation. When the farmer or gardener prays for the normal rainfall to come in its usual season, and when he prays for the rain to hold off during the normal season of dryness in order to accommodate the harvest, this is simply asking for this enforcement of normal natural function. When athletes pray that God would bless their game, allow them to play well, and keep them from harm, this is another example of asking for enforcement of normal natural processes. They are not asking to perform above their level of athletic skill. They are asking for God to allow and enable them to play their best, and they are in effect saying "May the best team win."

God can also respond with some special adjustment of the natural course of forces and events. When the hikers are hurrying home from their expedition and the clouds blow in dark and stormy, their prayer might be that God hold off on the rain until

---

18   "Come, My Soul, with Every Care" by John Newton, *Lutheran Service Book* #779, St. Louis: Concordia Publishing House, 2006.

they reach home. No miracle is requested or required, simply that God control the rain for the convenience of the hikers. When the basketball player stands at the free-throw line at the end of the game and prays: "God, help me to make this shot! I generally make 80% of my free-throw shots, and I am ready to attempt this one. But this one is important: this one decides whether my team wins or loses the game. Please help me, God, to make this shot properly and win the game for my team!", this asks not for some neutrality on the part of God, but for a special blessing that controls the outcome of the athletic contest. When soldiers go into battle, they do not pray that the best force may win, but they pray that God will intervene so that the battle tips in their favor, and that they survive and come back safely. They know that this means the other army has to lose, and that other soldiers have to die. They are not asking for some miracle on God's part. They are not asking that fire and brimstone fall from the sky, or that all the rifles of the enemy jam and cease to work. They are asking that the natural processes of battle work out, but that God tip the balance so that they gain the victory.

Prayers for medical healing usually are similar to the previous examples. The person might pray that his infection with the flu go its normal course, but that God enable him to endure the symptoms, and that he might recover in the normal fashion in as quick a time as possible. Or the person might ask for some extra help. "God, I have injured my ankle, and I am about to have an x-ray. Please arrange that my ankle is merely sprained, and that the bones are not broken so that it would require months to heal." This prayer asks for God to adjust the natural course for the benefit of the injured person.

Other prayers for healing can move into a third category of prayer, when disciples ask God for something contrary to the normal course of natural events. When a person has been diagnosed with cancer, they do not ask that God supervise the normal natural course, but that God intervene and change the natural course: make the cancer go away! This is quite legitimate for two reasons: God has asserted previously his interest in being of help to the person, and God has the power to control both the healthy and malignant cells acting within a person's body. This kind of prayer for healing is considered appropriate.

So we see that there are different kinds of petitions in prayers, which are related in different ways to what we understand about the aspects of probability in science. We can categorize three types of prayers: a) those that ask God to supervise the normal natural processes; b) those that ask God to intervene to control the normal natural processes for the benefit of the petitioner; and c) those that ask God to intervene with the normal natural processes in a way that contradicts them, for the benefit of the petitioner. This last kind is commonly understood as asking for a miracle, for an intervention by God that contradicts the normal natural processes. These three different types of prayer are related to the understanding of probability in connection with the normal processes of nature. Type (a) is not asking for anything that is improbable according to secular understanding of normal scientific processes. Type (b) is asking for something that is entirely within the normal range of probability of natural processes, but something that is not necessarily most likely, that is, that has a high probability. It may be asking for something that has a low probability, as we understand the normal processes of nature. But if this type of prayer receives a favorable response from God, this would not be labeled a miracle by the non-believer. The favorable response was improbable, against the calculated odds, but rare things do happen. There is a difference between rare and impossible. The religious believer would identify God's action of a favorable response as a miracle, but religious people tend to identify all of God's acts as gracious acts of mercy and as kind miracles performed in favor of his disciples. It is only type (c) that both groups, disciples and non-believers, would classify as miracles: when something happens that is, as far as we can tell, in contradiction to the normal natural processes. To the best of our understanding, the scientific probability of this event is zero.

It was stated in the previous chapters that the game of science studies the normal regular processes that God has set up to operate, and that the game of religion provides a method of addressing the irregular happenings, the special or miraculous events that occur. Religion provides a way to interpret when God causes something to happen that cannot be explained by science, because it is outside the normal processes, and outside the normal probabilities that govern normal natural processes. Note that the game of religion does not guarantee the ability to explain any action by God. Special events can be interpreted only in so far as they seem to be consistent with the purposes and promises of God as revealed in the sacred scriptures. The Christian Bible contains numerous examples where God chooses not to explain to his disciples what he causes in history. The book of Esther explains how God intervened in Persian history to save the Jews, but the book of Job refuses to provide any explanation for why God chose to allow the Devil to persecute Job and cause so much suffering. The book of Acts contains accounts of God miraculously saving some disciples from suffering and death, and accounts of God allowing some disciples to suffer martyrdom or great harm. If the game of religion is about the study of the supernatural, remember that humans cannot know anything about the supernatural world except as the supernatural world chooses to reveal something to those in the natural world. Thus religion is limited to what is revealed, and cannot reach beyond that to topics that are not revealed.

But those things that we can understand result from a combination of science and revelation, a combination of the study of the regular natural processes that God has designed, and of those things that God has revealed through sacred scripture. Religious believers are held accountable by God to perceive and take into account both of these sources of information. They are both acts of God. God has designed and created nature, and he has designed into it a sense of his power and his will in the way nature functions. And God has spoken by revelation by intervening in the natural world to convey additional information. God's will as unveiled by revelation does not contradict God's design of gravity, for example, nor God's design of the power of fire. If you stick your hand into a flame, God will cause it to burn. It does not matter how much God loves you; you are choosing to do something foolish and something against God's will when you deliberately stick your hand into a flame. God does not want your hand there; or else he would not have placed a flame there. You do not have to seek God's will in some special prayer. The presence of the flame functioning in connection with the regular processes of nature already exhibits God's will for that space.

But God can and sometimes does perform miracles, acts that are contrary to the normal processes of nature. With regard to fire, the notable story in the Christian Bible is the story in the book of Daniel of the three men who were forced into the furnace by King Nebuchadnezzar. These men were not disobeying God or displeasing God. They were seeking to obey the explicit instructions of God. They were forced into the blazing furnace by people acting in defiance of the God of the Old Testament. In that case, God chose to intervene, according to the story, and overrule the normal processes of nature: the men walked around in the fire and then came out unharmed. It is fair to assume that in this case the men had prayed to God that he would in some way save them from doom when the king had them arrested and sentenced. We would label this type of response to prayer as a type (c) response, because God acted in a miraculous way to preserve the men from harm contrary to all normal natural expectation.

But what happens when people act contrary to the will of God, and then pray for God's miraculous intervention to prevent them from suffering from normal natural consequences? I suggest that this should be understood as an error, or inappropriate behavior, that occurs when playing the game of religion. I call this irresponsible prayer. In discussing such a situation, we need only discuss the matter of religious believers. Since non-believers have no respect for God, there is no sense in which they can choose to act as following God's will. So even if at some point of danger they do suddenly turn in prayer toward God, that constitutes a complete reversal for their relationship with God, and that situation of a reversal from disbelief to faith is not the specific situation I want to discuss here. When someone turns to God with a plea for help after such an abrupt change in his attitude toward God, God can choose to respond either with mercy or by withholding mercy, however he chooses, in keeping with whatever greater plan he is performing. My interest at this point is discussing those people who are disciples of God, and who ignore God's will, but then turn and pray for release from the consequences. What should we expect of God in response to such cases?

To consider this, first let us examine a situation that is not obviously against God's will, and yet is viewed negatively by religious believers because it is not clearly aligned with God's will. This is when a disciple asks for a type (c) prayer, a special miracle on his behalf, that has no connection whatsoever with God's plan and God's direction for the believer's life. Take the example of when a believer asks God to cause him to win the lottery and gain the large financial payout, or to win while playing roulette, or to win the large-stake poker hand. It is true that God desires to help the disciple, and that God has the power to affect the outcome of the lottery or roulette wheel or poker game. But why should God cause one disciple to win the lottery and another disciple not to win? If there was some special situation connected to the petition, it might be justified. Perhaps an orphanage is on the verge of closing due to unexpected repair costs, and the lottery winnings will solve the problem and allow the good work of caring for the children to continue. But apart from such a special situation, we question the justification of such a petition. It is proper to ask God for help in continuing God's purposes. If God has caused a person to experience life, it makes sense to ask that God help continue that life, for example by healing cancer. But if God has not made any of his disciples rich, why should he act now to make one rich, and not others? Why does the disciple not pray that all the entrants win the lottery, or win the poker hand? But of course, if God acted in such a way, there would be no winner, for at best everyone would get back their cost of entrance in the lottery or the poker game. Here we find an example of a prayer request that we think may not be in accordance with God's desires.

    This doubt about the justification for the prayer becomes all the stronger if the person making the request has not done his part in trying to provide for his financial condition. When the person does not work well at a steady job, and neither budgets his finances nor saves for future needs, and then asks God to supply him extra money, we human beings consider such a prayer inappropriate. Note: I am not talking about a situation in which, through no fault of her own, a person ends up retired with no pension and almost no income, and prays to God for financial help. In such a case, when a person has tried all her life to carry out her responsibilities in caring for her family, or working for a living, and ends up through a set of circumstances in great need, it is considered appropriate to ask God for help, even financial help. But Christian, Jewish, and Muslim scriptures make it clear that

God expects each human to earn a proper living by some contribution to society, taking his place as a worker who contributes some benefit to society, and who draws out an income as some fair share of the blessings that God bestows on the society. The person who shuns participation in the community work effort does not deserve a share of the community bounty, and we believe—in accordance with the revealed scriptures—that he does not please God when he asks for such a share, or indeed asks for something greater than a proportional share. We believe that God has revealed a certain moral structure that teaches a relationship between contribution and benefit.

However, we also recognize that God often arranges things that are contrary to a simple contribution and benefit system. Some people work hard and honestly, and yet end up poor and needy. Some people do no work, but inherit large fortunes, or do win the lottery, or have oil discovered on their land. Others offer service to society, but are rewarded in disproportionate measures. For example, two people may go into musical careers. Both have equal talent and an equal work ethic, but one ends up old and poor, and the other was part of a band that sold millions of recordings and retires wealthy. The second was "lucky," while the first was not. Some people who are poor have lived irresponsible lives, and some have tried to live honest and responsible lives. Some people who are rich fall into the category that we call responsible, and some wealthy people do not. Some rich people use their wealth and position to try to contribute to the benefit of society. This does not require that they give up their wealth; it only means that they try to serve. Perhaps a wealthy person goes into politics, devoting his time to trying to solve the problems of his community. Other people who are rich use their wealth only to entertain themselves, and do not operate from any sense of responsibility to find and accomplish God's will in the greater society.

From what we learn of God in the revealed scriptures, we expect that God will receive the prayers of responsible people and irresponsible people in different ways. We expect that God will look with favor upon the prayer of a needy responsible person, and God will arrange some way of helping that person—possibly through the aid of other disciples. We expect that God will look with favor upon the prayer of the responsible rich person and provide help, to whatever measure it is in accord with God's greater plan. If the rich but responsible person's yacht burns and the insurance company tries to avoid payment, it is fair for the rich person to ask God for help in arranging the outcome of the legal process. If the irresponsible rich person asks for help, we expect that, unless God has other special reasons, God will not look with favor upon his prayer. For example, if the rich insurance man tries to avoid paying on the policy for his neighbor's burned yacht, and he asks God to help him win in the legal process and avoid having to pay, we expect God to turn down such a request.

So we understand that, according to the revealed scriptures of the major religions, when a person brings a petition to God in prayer, God's response is somewhat conditioned on the behavior of the petitioner. There are times when asking for certain things from God is contrary to the plan that God has previously revealed in his sacred scripture. We can classify many of these situations as irresponsible prayer. God has revealed his grace and love toward his disciples, and God wants to work with his people. But God expects his people to work with him also in his plans. When a person works against God's plans, his request for help is offensive to God.

So prayer is a legitimate exercise of religion that may ask God to coordinate the normal processes of nature in favor of the petitioner, or that may ask God to intervene with a miracle in contradiction to the natural processes. This does not mix or confuse the rules of the two games of science and religion. Prayer is a deliberate appeal to the supernatural to intervene in the natural world on behalf of the believer. This intervention may contradict the expectations of science, but the prayer does not contradict the rules of playing the game of science. The prayer is an appeal to the other game.

Chapter Ten: Comments on the Genesis 1 Creation Story

Over the last century in the United States much of the tension in the relationship between science and religion has occurred in connection with the understanding of the creation story in Genesis 1 of the Bible. Therefore it may be helpful to give a little more attention to this story in light of what has been said in this book about keeping distinct the games of science and religion. The study of science has developed so that it has much to say in regard to the formation of our planet and its life, and what this science says does not seem to coincide with the opening story of God's creation of the world as narrated in the sacred scriptures of Christians and Jews. This creates difficulty when a person tries to integrate these two sources of information. This difficulty is greatly increased when one confuses the two different games of science and religion, and it is greatly decreased when one makes the effort to keep the games separate and respect their different rules. What will be said in this chapter will not resolve all of this difficulty, because, as has been pointed out in Chapter Two, we do not know everything. We do not know everything in the domain of science, and we do not know everything in the domain of religion. We are still learning more about the sciences of geology, biology, and cosmology, and we are still learning more from archaeology, history, and linguistics that affect our interpretation of the sacred scriptures in religion.

In the first chapter of the Bible, we find a narration of the creation of the world by God. The story is described as spanning six days. The act of creating the world ends with the creation of the human being, both male and female, and the assignment to humans of the task of management over the rest of the world. Then God "rests" from his work of creation. There are numerous questions that come to mind with the most basic inspection of this story. If the sun is not created until day number four, what is marking time for the first three days? If there are no humans in existence until the end of this week, who is observing this event, and who is describing it? The obvious answer to this question would seem to be that God is revealing this to humans. But in the narration, God does not speak in first person (e.g., "On the first day I created light" etc.); someone else is speaking about God in the third person. The use of the six-day week is curious. Why should the Almighty God need six days to construct the parts of creation? Why not carry out this task in six hours, or six minutes? Why does the time span of seven days correspond to the normal seven-day week of later human history, which is presumably based on the division of the cycles of the moon? When God created all the vegetation of the Earth on day three, including their process of replication by seeds, are we supposed to envision those first plants surging into existence and rising to their full growth in a miraculously rapid time frame? In other words, do the first oak trees grow up over decades or within minutes? When God created the creatures of the sea and land, he commanded them to multiply and fill the Earth. So did God create only one pair of each, or did he create many multitudes of animals, but a number still below the quantity needed to fill the seas and the earth? In the light of this question, when God created humans, did he create only one couple, or several? It is common in the rest of the Bible to speak of a group of people in a collective sense as represented by one ancestor. God often speaks of his people by speaking symbolically of "Jacob" or "Israel" or "Ephraim" (for example see Hosea chapters 6-8, or Isaiah 41:8-16; 42:18-43:7; 43:22-44:5; 45:4, or Psalms 73:1; 78:5, 21, and 71; 80:1-2; 81; 105:37-38; 114:1-2; 118; 121; 124; 129; 130:8; or 135). He also personifies his people as a single person who serves as his bride, and calls her by the male name "Israel." In the Psalms we often encounter the case of an individual speaking in first person voice addressing God, but modern scholarship suggests we should frequently understand that person as the king, speaking as God's ideal disciple, representing all of his fellow subjects. So the label "Adam" could be part of this pattern to represent all humans. The usual understanding that Adam is one person is taken from the following story in Genesis 2, where we read of one human male, and then a cloning operation to give us one human female. But even this interpretation is complicated by the fact that the story does not give a name to the human, but simply calls him "the man." The Hebrew word "Adam" is the word that means "man." Actually this would be better translated into modern American English as "the human"; notice that in Genesis 1:27 the creation of "man" is the creation of mankind, and includes both male and female sexes. On the seventh day God "rested" from his work of creation. Why? Was God tired? Did this creative work take a lot of energy out of God? That seems unlikely. Are we to understand that God stopped working for a while? That would seem to conflict with many other passages that occur later in the Bible which describe God as one who does not sleep or slumber, who never tires, and who is not bound by human conceptions of time (compare Psalm 121:4, or Isaiah 40:28). The Hebrew word used is the verb that lies behind the word "sabbath," and it means to take a break, not to take a nap. God "rested from his work" of creating things. He did not need to rest his body or mind and recover his energy; he took a break from what he was doing, and presumably gave his attention to something else. But again, this shift of action or attention fits what God is doing into the pattern of the ancient Israelite work week, and this seems like an intentional narrative convention, rather than a simple fact of history.

It is furthermore obvious to any reader that there are inconsistencies between the stories of Genesis 1 and Genesis 2. In Genesis 1 God created all the plants on day three; then he created all the animals on days five and six, and then he created the male and female humans together. In Genesis 2 God created the male human first, when "no plant of the field had yet sprung up" (Genesis 2:5). Then God created the plants, then the animals, and last of all God came to the business of the creation of the female human. For some reason it has become a significant thing for modern scholars to point out these differences, as if this was a modern discovery that throws new light on the understanding of the Biblical story. But these differences are immediately obvious to any reader who compares these two chapters, and do not require any advanced training to be recognized. Surely these differences were also clear to the ancient Israelite readers, young and old, of centuries ago. The fact that the Biblical text makes no effort to harmonize these details, or even to discuss them anywhere else in the Biblical story, offers very important information to us regarding the understanding of the intention of the original writer, and regarding the understanding of the original audience. One way to understand this is to compare this situation to viewing the video replays of a certain event in a football game. From one camera angle it appears the player was out of bounds, but from another camera angle it is clear that he was in bounds. In a similar way these two texts discuss the creation of the world from "different angles," and it is neither necessary nor legitimate to force them to contradict each other.

In addition to these questions, there is the question of the relationship of these first chapters of the Bible to the other descriptions of God's creation of the world in other parts of the Bible. Psalm 104 is recognized as a poetic reflection of the six-day creation story in Genesis 1. But in the poetic retelling of the story, God is pictured much more in the image of a human craftsman erecting a structure. God "set the earth on its foundations"; he "stretches out the heavens like a tent, and lays the beams of his upper chambers on their waters" (Psalm 104:5, and verses 2- 3). The mention of the waters in this verse refers to "upper chambers" and must mean the waters above the sky mentioned in Genesis 1:7. So God stretches out the sky like a cloth canopy to keep out the material above, and then above this canopy God fastens the "beams of his upper chambers" in the watery superstructure so as to provide a secure place for him to sit above the Earth. This corresponds to Job 26:11 (below) where pillars are described as holding up the sky, as would be the case in a human building with pillars to hold up the ceiling. Job 9:6 (below) mentions other pillars that hold up the earth, as in Psalm 75:3 (below). These pillars are equivalent to the "foundations" that support the surface of the earth, and which can be shaken or toppled, if God allows, or held firm, if God chooses to be merciful. Consider some of the many passages that refer to these foundations and pillars.

[Job 26:7-11]
[God] spreads out the northern skies over empty space;
    he suspends the earth over nothing.
8 He wraps up the waters in his clouds,
    yet the clouds do not burst under their weight.
9 He covers the face of the full moon,
    spreading his clouds over it.
10 He marks out the horizon on the face of the waters
    for a boundary between light and darkness.
11 The pillars of the heavens quake,
    aghast at his rebuke.

[Job 9:6]
He shakes the earth from its place
    and makes its pillars tremble.

[Ps. 75:3]
When the earth and all its people quake,
>   it is I who hold its pillars firm.

[Ps. 82:5]
They [sinners] know nothing, they understand nothing.
>   They walk about in darkness;
>   all the foundations of the earth are shaken.

[Ps. 102:25]
In the beginning you laid the foundations of the earth,
>   and the heavens are the work of your hands.

[Ps. 104:5]
He set the earth on its foundations;
>   it can never be moved.

[Prov. 3:19]
By wisdom the LORD laid the earth's foundations,
>   by understanding he set the heavens in place.

[Prov. 8:27-29]
I [Lady Wisdom] was there when he [God] set the heavens in place,
>   when he marked out the horizon on the face of the deep,
> 28 when he established the clouds above
>   and fixed securely the fountains of the deep,
> 29 when he gave the sea its boundary
>   so the waters would not overstep his command,
>   and when he marked out the foundations of the earth.

[Is. 48:13]
My own hand laid the foundations of the earth,
>   and my right hand spread out the heavens.

[Is. 51:13]
... that you forget the Lord your Maker,
>who stretched out the heavens
>and laid the foundations of the earth.

[Is. 51:16]
I who set the heavens in place,
>who laid the foundations of the earth.

In the following passage Isaiah describes God's coming punishment of the destruction of the world; but note how this mentions the shaking of the pillars of creation such that the Earth shakes and falls:

[Is. 24:18-20]
The floodgates of the heavens are opened,
>the foundations of the earth shake.[19]

19 The earth is broken up,
>the earth is split asunder,
>the earth is thoroughly shaken.

20 The earth reels like a drunkard,
>it sways like a hut in the wind;
>so heavy upon it is the guilt of its rebellion
>that it falls—never to rise again.

[Ps. 93:1]
The world is firmly established;
>it cannot be moved.

---

[19] The word translated "shake" in verse 18 and 19 would be better translated "collapse." See Paul Puffe, "מוט Means 'Collapse,' Not 'Be Shaken,'" *Concordia Theological Journal* 6:2 2019, 77-92.

[Ps. 96:10]
Say among the nations, "The LORD reigns."
    The world is firmly established, it cannot be moved.

The prophet Isaiah likes to use the image of a tent-erector for the picture of God's act of creating the world. God stretches out the tent above, and unrolls the rug of the earth below:

[Is. 42:5]
This is what God the LORD says—
    he who created the heavens and stretched them out,
        who spread out the earth and all that comes out of it,
    who gives breath to its people,
        and life to those who walk on it: ...

[Is. 40:22]
He stretches out the heavens like a canopy,
    and spreads them out like a tent to live in.

[Is. 34:11]
All the stars of the heavens will be dissolved
    and the sky rolled up like a scroll;
all the starry host will fall
    like withered leaves from the vine,
    like shriveled figs from the fig tree.

[Is. 64:1]
Oh, that you would rend the heavens and come down,
    that the mountains would tremble before you!
    [note: "rend" or rip the sky like a cloth tent cover]

All of these passages raise the question of which picture of creation should be understood as definitive. Did God create the world like the erection of a tent, or like the construction of a vast building, or like the creation of a diorama? Genesis 1 is strongly reminiscent of modern-day humans who create full three-dimensional landscapes for their model train sets. First you must make some space in the house or garage, then you set up the table, then you lay out the terrain. After all of this you place the track and set in place the buildings and all the little figurines. After six "weeks" of detailed work, you plug in the transformer, sit back and run your train set. And if your train goes off the track and damages your diorama, you have to get up and return to work fixing your creation, as God has had to do now and then in Biblical history.

Some people argue that Genesis 1 should be taken as the definitive description of creation, and the other passages should be understood as figurative pictures, only meant to illustrate the powerful creative ability of God. Genesis 1 is assigned this priority because of its placement at the beginning of the Bible. It is argued that the "simple, literal" reading of the text should imply that this was intended to be the definitive story. However, this argument begs the question of what is the "literal" understanding of the sacred text. It asserts that one part of the Bible is meant to be "historical," and other parts are meant to be figurative. It does not establish that Genesis 1 was not meant to be understood as a figurative description in the same way as the other passages. That assertion does not prove the interpretation is correct, it merely offers one possible interpretation.

For some people, the creation descriptions in Proverbs and Isaiah are not seen as in conflict with modern science and cause no problem for religious understanding, but they find it disturbing that the interpretation of the Genesis 1 creation story seems to be in conflict with the information presented from the discoveries of modern science. This leads to their insistence that we must cross between the games of science and religion in order to harmonize these two sets of information. Great effort is expended to explain how an interpretation of Genesis 1 might be made compatible with the data of modern science. In this particular interpretation, often labeled as "creationism," [20] it is insisted that the six days of Genesis 1 must signify a time span corresponding to six normal days of our human experience. These interpreters often also reason that the information given about the life spans of the ancient patriarchs in Genesis 5 and Genesis 10 should be understood to teach that the Earth is only a few thousand years old. A calculation of the life spans in those chapters, when correlated with the later events of the history of the Old Testament that we can place within our calendar dating system, gives the conclusion that the world is about 6 thousand years old. Some of these interpreters are willing to allow for some measure of uncertainty in these dating calculations, so that the Earth is perhaps 10 thousand years old. But this is in contrast to the results of modern science, which tell us, according to the latest calculations, that the universe is 14 billion years old and the planet Earth is 4.5 billion years old. These devotees of creationism give priority to the calculation taken from the Biblical data, and thus it becomes necessary for these interpreters to explain how the scientific data is either in error, or is being wrongly interpreted; and how the scientific data should be interpreted to correspond with the 10 thousand-year-old age of

---

20   This view labeled "creationism" is the belief (-ism) in a certain interpretation of creation, that God used six 24-hour long days to make the world. But it becomes confusing when people further use the label "creationist" to identify these adherents, because the label "creationist" is also used to identify people who believe in a divine creation as opposed to a merely natural evolution of the cosmos and biology, but who don't insist on a 6 day = 144 hour time frame. One can contrast the views of the members of the *Answers In Genesis* organization with those of *The BioLogos Foundation*, or *The Faraday Institute for Science and Religion*.

the universe and the specific six days of creative activity on the part of God.

One of the big problems with this suggested interpretation, which most of its adherents fail to recognize, is that when they mix the games of science and religion they are very arbitrary about what is miraculous and what should follow the accepted rules of science. Most of these interpreters seem to assume that when God created the world, the laws of physics and chemistry must have been created in the first instant, and then everything else that God did during creation had to somehow work within, or at least alongside, those laws. Thus we have the existence of space and time in effect at the beginning of the first day of creation, and the following six days simply describe the additional things that God created on those days (light, atmosphere, dry land and plants, etc.). There is no need for this assumption of the prior existence or creation of the laws of physics. This is forcing certain data from the game of science to govern the interpretation of the description of a miracle in the sacred scripture that reveals information about the supernatural in the game of religion. Surely everyone will agree that the act of creation is a miracle. Sometimes a miracle consists of God overriding the laws of probability to ensure a desired outcome, but more often a miracle consists of God acting outside of and in defiance of the normal processes of nature that he has designed and put into effect. If creation is a miracle performed by God, there is no reason to expect any part of it to conform to any of the normal rules of physics, chemistry, or biology.

But actually the problem with this one interpretation of Genesis 1 is even worse. The debate inspired by this interpretation dates back to the nineteenth century, but really not before that. In our time (the twenty-first century) it seems like this has been the eternal debate between religious believers and non-believers because to most modern people this is so "old."

But from the standpoint of academic history this debate is relatively recent or "new." This debate involves trying to interpret the Biblical creation story as a historical event only in relationship to the status of scientific understanding that was current in the nineteenth century. It was in the nineteenth century that so many areas of scientific study were developed. A few questions preceded this time, such as the matter of geocentrism or heliocentrism (does the sun circle the Earth or the Earth circle the sun?), but it is in the nineteenth century when this tension between science and religion began to be important. This tension became acute when Charles Darwin's theory of natural selection gave great impetus to the idea of biological evolution, in connection with new understandings resulting from other areas of research concerning geology, archaeology, social history, and paleo-history. For example, prior to the nineteenth century there were no such things as dinosaurs. Fossil bones had been found previously, but they were not recognized as the remains of creatures that lived millions of years earlier. In the nineteenth century the combination of information in geology and biology began to create a dating scheme for ancient ages. The entire science of finding, categorizing, and speculating about dinosaurs has come into existence only in the last two hundred years. Only since the nineteenth century has there been any question of the relationship of dinosaurs to the Biblical story. Until that time no one had to debate whether or not dinosaurs were on Noah's Ark because no one had yet conceived of dinosaurs. All of this new information in the game of science contrasted with the interpretation of the Biblical creation story that had been common in Europe, which taught that the world was only a few thousand years old and that all types of creatures had been created separately by God at the beginning, relatively recently (compared to the alternative idea of a planet over 4 billion years old), and that none had been allowed to die out. This new information created a great tension between these two sources of truth.

The understanding of science that was used as the basis for discussing this tension can be labeled a Newtonian view of science because it was built on the understanding of physics that resulted from the work of Isaac Newton (1642-1726, the seventeenth-eighteenth centuries). Newton had produced mathematical formulas that described the motion of particles. These formulas applied to both large particles such as planets in orbit around the sun, and small particles such as atomic particles (atomic particles such as protons and electrons were not known by Newton, but his laws of motion were used in the following two centuries to investigate their characteristics). In the nineteenth century these Newtonian laws of motion were considered to be fundamental, absolute laws that governed the functioning of everything in the universe. It was only in the twentieth century that scientists came to understand the limitations of these formulas in regard to physical processes operating in large or small extremes such as astronomy or atomic particles. The sciences of astronomy, cosmology, and atomic physics have advanced greatly since the nineteenth century, and Newtonian physics has been replaced by more intricate formulas created by more recent physicists such as Albert Einstein and Werner Heisenberg. In the nineteenth and early twentieth centuries those who sought to find some harmony between the traditional religious view that God had created the world in six days only a few thousand years ago, and the new scientific information that indicated a much longer history for geology, biology, and astronomy, used calculations based on Newtonian formulas in an attempt to explain the relationship between the two descriptions of phenomena. Today some adherents of this method of interpretation try to incorporate the newer scientific formulas for physics resulting from Einstein's theory of relativity in a similar attempt to find a possible harmonization between the information from science and religion. But both the older and the more recent groups of these interpreters still make the assumption that any reconciliation of the Genesis account of creation must fit within what we understand about the laws of physics and biology. The fundamental assumption is that when God began to create the universe, he first created the laws of physics, and everything thereafter had to fit logically into the functioning of those laws.

But there is no need to begin with the assumption that the laws of physics, chemistry, or biology apply to the description of the creation story described in Genesis 1. If you shake off that traditional scientific framework as a starting point, the possible interpretation of the six days of creation in Genesis 1 can be viewed very differently. Let me offer an alternative understanding of the creation story of Genesis 1 that I have conceived. I think this alternative understanding is ultimately not fully suitable, but it is useful to illustrate the need to detach the discussion from that frozen Newtonian scientific framework.

In the beginning of the Genesis 1 creation story (Gen. 1:1) we have only God, and nothing has yet been created. If nothing has been created, there is no such thing as time and space. The first thing that God does is create time. He does this by creating light and separating it from darkness. But God does not name these two new features "light" and "darkness," rather he names them "day" and "night"[21] (Gen. 1:3). The focus is not on the phenomenon of light or radiation, and all the physics we might

---

21   I am indebted to this observation from John Walton in his book *The Lost World of Genesis One*, (Downers Grove, IL: InterVarsity Press, 2009), pages 53-55. This observation led to my contemplation of the rest of the idea.

want to associate with that. The focus is on the fact that God created change: something was different, and it varied. Where nothing changes, there is no passage of time. If everything stays the same, there is no such thing as time. The only way we have of observing time, and of measuring time, is by measuring the change in something. Be careful here: from science we are used to speaking about the space-time framework. But this passage of Genesis is only about the creation of time. You should not associate the creation of space with day one.

For the creation of space is the work of day two. You have a homogeneous universe with no difference in any feature in any direction; actually you can't even speak of direction. Now God creates a difference. He separates the waters above from the waters below by the creation of a thing, a divider (Gen. 1:7). This divider is the atmospheric dome that covers the land (which does not yet exist), but the atmosphere is not the main point. The point is the separation of space. The rules of physics do not apply to the miracle of creation. God can create time on day one and still not create space until day two. God creates both, but in a sequence. There was no such thing as "here" or "there" prior to day two. Using the language and cultural concepts of Iron Age Hebrew, the Bible describes for us the creation of space with different locations, thus giving us our three normal dimensions of length, width, and height.

You must not presume other physical laws are yet in effect. On day three we can conceive of the creation of gravity. On day three Genesis describes the exposure of the dry land, and the subsequent creation and growth of land plant life (Gen. 1:9, 12). Our usual picturing of this event is to envision the land masses of the continents rising up out of the sea. But that seems to differ from the conception of the ancient Israelites. In Psalm 104:5-9, the psalm that parallels Genesis 1 in the description of the six day creation account, it is not the land that moves up, but the water that moves down. The water below the divider initially covers the land. But something causes the water to move, to drain off the land. And something causes the water to be locked into this new position so that it does not flow up over the land.

[Psalm 104:5-9]
He set the earth on its foundations;
    it can never be moved.
6 You covered it with the deep as with a garment;
    the waters stood above the mountains.
7 But at your rebuke the waters fled,
    at the sound of your thunder they took to flight;
8 they flowed over the mountains,
    they went down into the valleys,
    to the place you assigned for them.
9 You set a boundary they cannot cross;
    never again will they cover the earth.

This description is consistent with other passages in the Bible that often refer to God's control over nature in terms of his locking the sea into its basin, including Gen. 1:9.

[Gen. 1:9]
And God said, "Let the water under the sky be gathered to one place, and let dry ground appear."

[Job 38:10–11]
Who shut up the sea behind doors
    when it burst forth from the womb,
9 when I made the clouds its garment
    and wrapped it in thick darkness,
10 when I fixed limits for it
    and set its doors and bars in place,

ⁱ¹ when I said, "This far you may come and no farther;
      here is where your proud waves halt"?

[Psalm 33:7]
By the word of the Lord were the heavens made,
     their starry host by the breath of his mouth.
⁷ He gathers the waters of the sea into jars;
     he puts the deep into storehouses.

[Jer. 5:22]
I made the sand a boundary for the sea,
     an everlasting barrier it cannot cross.
The waves may roll, but they cannot prevail;
     they may roar, but they cannot cross it.

What causes water to come down off the land and stay within its basin? This is what we today call gravity.

So on the first three days of the miracle of creation, translating into terms of our modern science, God creates time, then space, and then gravity. Also on day three God goes on to create plant biology.

Then on day four (Gen. 1:14), God gets around to creating the universe, in terms of sun, stars, and planets. This is so completely opposite of our conception of science that it is difficult to picture this. We insist that we have to begin with the creation of the universe of stars and stellar dust, then the collection of dust into planets, and then the creation of plants and other things on the planet. But Genesis 1 does not speak in terms of our modern scientific worldview at all. God is doing a miracle, and he can do it any way he desires. God can create time, space, and gravity, and toss in plants, without having yet created a sun or a solar system. God can have gravity drain the water off the surface of the Earth, an Earth that miraculously exists without the creation yet of the solar system, without the orbiting of the planets, without the shining of the sun. The Biblical text felt no need to speak to our concerns about planets and solar systems, because it was not speaking to people in the original audience who had any knowledge or conception of such things. Just as you don't describe differential calculus to a kindergartener, you don't describe planetary science to an ancient Israelite.

Day five (Gen. 1:20-23) relates the creation of aquatic life and bird life, and day six (Gen. 1:24-25) continues with the creation of land life. Day six of course culminates in the creation of human life (Gen.1:26-30). Only when we arrive at day seven are we justified in making the assumption that the normal laws of physics, chemistry, and biology are in existence and in effect (because God "rested" or ceased from further creation). Until that time, nothing in the Biblical account needs to conform to any part of what we think of as modern science.

This also raises the question of how the timespan of creation compares with modern time. Of course there is the very familiar passage of Psalm 90:4 and 2 Peter 3:8 that "with the LORD a day is like a thousand years, and a thousand years are like a day." This suggests that we should be careful about trying to hold God accountable to any of our conceptions of time. But look back inside the days of creation for a moment. When God commanded the dry land to bring forth vegetation, how long did it take the first oak and redwood trees to grow? Did they follow the normal biological process, and take the normal amount of time, as measured against the human life span? Or did they just shoot up into maturity? When God commanded the fish and the animals to multiply and fill the sea and the land, how long did this replication take? Was any biological process happening at the normal rate that we experience today? If anyone had been there with a stopwatch, how long would these events have taken? It appears that it is improper to attempt to apply modern scientific concepts of time to the creation miracle. This would be all the more true for any consideration of day four, the creation of the stars, sun, moon, and other features of the universe.

And yet, all of this is still to think of time as some sort of scientific constant, and to insist that our understanding of the creation miracle fit into this scientific framework. But current science denies this necessity. Current science, since the conception of general relativity, states that the experience of time is not a constant. The closer one travels to the speed of light inside our universe, the slower time passes for him, relative to the passage of time for those in the universe not traveling at that speed. That means that time is measured differently for different people. They do not experience the same amount of time. Modern calculation of the time dilation effect suggests that there is a difference in the experience of time just from experiencing even the slightest difference in acceleration. Even living at different altitudes on Earth will involve slightly different forces of gravity, and this will affect the experience of time.[22] Time is not a constant. Nor is the speed of light an ultimate limit. Though we are used to thinking of it as a limit, current theory about the beginning of the universe holds that after the Big Bang the universe expanded faster than the speed of light. That is why astronomers can observe light rays coming from objects 46 billion light years away, even though the universe is only 14 billion years old. It didn't take 46 billion years for that light to travel this far because the universe is not the same size it was 14 billion years

---

[22] "Relativity Affects Everyday Life," *Austin-America Statesman*, 25 September 2010; reporting on an article by C. W. Chou, D. B. Hume, T. Rosenband, and D. J. Wineland, "Optical Clocks and Relativity" in *Science* (24 September 2010: 1630-1633).

ago. These conclusions and theories from modern science about space and time, and about the theoretical moment of creation called the Big Bang, are actually fairly difficult to comprehend. For example, the Big Bang did not involve the explosion of matter to fill the universe. Prior to the Big Bang, there was no space and time. The Big Bang represents the coming into existence of space and time. If these are the kinds of concepts modern physicists have to work with, it makes no sense to shift to the game of religion and demand that supernatural miracles must conform to some kind of physics, either Newtonian physics or Einstein's relativistic physics.

This way of looking at the six days of creation in Genesis offers a way to break through the locked focus of the traditional Creationism argument about how to interpret Genesis 1 as fitting into a scientific framework. I stated above, however, that ultimately I do not think this way of understanding Genesis 1 is satisfactory. My reason for this comes from the game of religion, not from the game of science.

The game of religion is all about interpreting revelation from the supernatural world. When that revelation consists of sacred scripture, it is necessary to follow the proper rules to interpret the scripture. Those rules call for making the best attempt possible to uncover the original purpose and meaning of the original writer, taking into account the probable original understanding of the original audience. The interpretation must be properly grounded in all that we know of the history associated with the sacred literature.

For the interpretation of the Bible concerning Genesis 1 and related passages, that means we need to take account of the conceptions of cosmology held by the ancient Israelites, and of the conceptions of religion and religious topics held by the ancient Israelites. When God provided that part of the Bible, he did not cause it to be written in twenty-first century American English. It was written in Iron Age Hebrew.[23] We have to translate from the language of Hebrew to English. We also have to translate from the culture of the Iron Age to the modern age. When the Biblical writer speaks about a dog, we cannot import our understanding of "man's best friend," but we must learn, as best we can, what the ancient Middle Eastern peoples thought about dogs. Then we must test that to see if it makes sense in the context of the Biblical statements. In the Middle East dogs are generally considered useful but dirty farm animals. They are not usually treated as house pets, as Americans do. This brings a different understanding to what is meant when, for example, Jesus states that is not right to give the children's bread to the dogs (Matthew 15:26-27). Another example is that when the Bible talks about government, we need to translate into modern terms. The Bible refers to respect in one's interaction with kings. By interpretation we take that to mean respect in one's interaction with human governmental authorities. Therefore we apply these principles to our interaction with appointed and elected city officials, who are nowhere close to the position of ancient kings. When the Bible talks about ancient social structure, we need to translate that into modern terms. The Bible talks about the rules regarding slaves, and about the treatment of the poor. We don't have slaves today. But we interpret those passages to speak about how we are to treat people who work for us, and who are lower on the social ladder than we are. So we apply the principles to how we treat our employees and those of lower economic status. Our employees are not slaves. But we will not find passages in the

---

23 Some parts of the Bible are in the earlier form of the language of Bronze Age Hebrew, which raises interesting questions about the history of the Biblical text, but for the most part we cannot answer those questions.

Bible giving instructions about union and non-union employees, laborers and exempt administrative personnel, and all the different types of economic labels that we have today. We have to translate.

This same principle applies to the ancient Israelite understanding of science and cosmology. We cannot expect the Bible to talk to us today in terms of planets, hemispheres, or meteorites. If God had spoken English to the ancient Israelites, that would not be communication. If God had spoken to the ancient Israelites about solar flares disrupting radio communications, about the distance to the American continents, or about electricity, that would not be communication. If God had wanted to explain something about these features, God is certainly capable of doing so. But the text would make clear that God was intending to explain some special, otherwise unknown, feature. God could have used and defined an English word, and God could have discussed the distance to the American continents. But it is obvious that God did not do this. So when we read the words addressed to the ancient Israelites, we need to expect that they will be phrased in terms that were understood by those Israelites. That means that God chose to limit himself to speak in terms of the science of their day. He did not choose to speak about scientific information that humans would discover in later times. When God talks about the creatures of the sea, he leaves it at the level of knowledge of aquatic life familiar to the ancient Israelites. He does not distinguish a dolphin from a fish or a shark, and he does not talk about squid or penguins. When God talks about geography, everything is organized around the land of Palestine. The Hebrew word for "west" is "seaward" because everyone knew where the Mediterranean Sea was. The "far-off islands" were the nations located far in the west of the Mediterranean. The Israelites had heard of Cyprus and Crete and Rhodes and Sardinia and Malta, but for all practical purposes none of them had ever traveled there. The great powers of Mesopotamia threatened to invade from the "north," not the east, because that is the way that all travel came to Palestine around the Fertile Crescent. All geographical terms are related to the Israelites' geographical sense of the world.

This also applies to their sense of cosmology, which means their conception of the shape of the universe. When God spoke to the ancient Israelites about the world, he spoke in terms of their ancient Middle Eastern conceptions. The world did not consist of a global planet for them. It consisted of a huge disk of dry land, held up from falling into the sea by great mountain pillars, with a great dome overhead keeping out the rest of the water that filled the universe above and below the Earth. Rain, snow, and hail occasionally fell through the dome from the sky. (Job 38:22 speaks in terms of storage bins for snow and hailstones; Genesis 7:11 and 8:2 speak in terms of "floodgates of the heavens" being opened. See also Deuteronomy 28:12, 2 Kings 7:2, Psalm 78:23, Psalm 135:7, Isaiah 24:18, Malachi 3:10.) This was the world as they understood it. When God describes the creation of the world, he describes it in terms that the ancient Israelites understood. It made sense to them; they understood the message. The world did not consist of all kinds of different supernatural gods that lived in and controlled different domains, such as the sea god, the wind-and-rain god, the god of the underworld, or the god of some volcano. The Israelite God, their God, the one God, had made all of it, and controlled all of it. And God had the power to unmake it. He did so at the time of Noah. He let the water rise above its level, and he opened the windows of heaven, and the waters undid the separation of creation day number three. God undid all the creation of life on days five and six, except for the few survivors he placed in the ark. Then, in Genesis 9, God recreated the world and gave it the same instructions he had given previously. When we read the creation account in Genesis, we must interpret it according to our best understanding of the ancient Israelites' conceptions of cosmology, and then translate that into our modern, completely different understandings of science and cosmology. So we have one God, who created the universe, the Earth, and all life on this planet. This God created humans and placed them in a management role over the rest of the planet.

This same necessity applies to the Israelites' concept of religious topics. We must seek to understand what connotations they thought were involved in the use of different phrases and ideas. What did the parts of the creation story signify for them in terms of important religious topics? The use of the seven day week is certainly tied to this religious understanding. We are all the more certain of that when we observe how the length of a week is used in the religious literature of the Israelites' neighbors. In the Ugaritic documents, when Baal builds his temple, it takes one standard week of work. Was there a common religious understanding in the ancient Middle East that a god would take a week of work to construct his temple? Some scholars think so.[24] It is quite likely that ancient temples were designed with significant symbolism related to ancient cosmology. The temple that Solomon built was adorned with two great pillars labeled "Boaz" and "Jakin." These may symbolize the pillars that held the world up out of the water. In front of the temple was a great tank of water. This probably represented the sea as a tame body of water under God's control. The interior walls were decorated

---

24 See John Walton, *The Lost World of Genesis One: Ancient Cosmology and the Origins Debate* (Downers Grove, IL: Intervarsity Press, 2009); and Margaret Barker, *Creation: A Biblical Vision for the Environment* (New York: T & T Clark International, 2010).

with plant images, representing the beautiful garden God had made on Earth. The walls also had images of "cherubim" or angelic agents of God, probably symbolizing his control reaching into the entire world, as well as symbolizing the angels surrounding God in heaven. 1 Kings 6-7 describe Solomon's temple; Isaiah 6 and 1 Kings 22:19 mention the angelic host. If the temple symbolized God's creation of the universe, then to what extent in the minds of the Israelites was the description of creation a symbolic statement about God's temple?

The structure of the creation story in Genesis 1 has an evident plan to it. In the beginning the earth is "formless" and "empty." In the first three days God *forms* the parts of the world. In the second three days he *fills* the structured world with active agents. The active agents correspond to the parts that were formed. On the first day God creates the phenomenon of light, and on the fourth day (the first day of the second set of three) God fills the heavens by creating agents to shine and give light: the sun, moon, and stars. On the second day God divides between sky and sea, and on the corresponding fifth day he creates agents to fill the sky and sea: birds and fish. On the third day God creates the dry land and plants, and on the corresponding sixth day he fills the land by creating land animals to eat the plants, and humans to eat the plants and to tend the plants. This structured shape of the narration of the creation story is so clear and so simple that it raises the question of whether anyone is justified in trying to find any deeper meaning in it. The narrative paints a very clear picture of God creating the universe as a three-dimensional diorama (remember what was suggested above about humans creating a model train set). Did the ancient Israelites find any significance in noting the clashes between Genesis 1 and Genesis 2, or between Genesis 1 and Isaiah's tent-erection imagery? Or would they have argued that such contrasts are missing the underlying religious points about

God's control and God's intended purposes? How should these passages about creation guide our interpretation of the passages that forecast God's destruction of the world in judgment against sin, and God's promises to build a new heaven and earth in the future? The purpose of these passages is to reveal something about theology from the supernatural world, not to discuss history or science.

We should also be cautious about discussing the concept of science in connection with the ancient world, and thus in connection with the Jewish/Christian Bible. The concept of science as I have defined it in this book is built on the clear distinction between the natural and the supernatural as distinguished in the evolution of European scholarship after the Reformation. The ancient world also had a distinction between what was natural and what was supernatural. But the knowledge of the natural world in ancient civilization was so much more limited than our modern body of science that the ancient people most likely would have classified many things that we consider natural as part of the domain of the supernatural. Today we don't immediately associate phenomena such as lightning, volcanoes, sexual reproduction, or diseases, with gods or the supernatural, but the people of the ancient Israelite world probably did. In the Bible we read about the danger of evil spirits that might enter into a person's body and cause harm, as well as cause strange behavior. Today we speak about viruses that enter into a human body and cause harm, and possibly cause strange behavior. Where we today would distinguish between problems caused by a virus, problems caused by chemical issues such as epilepsy, and problems caused by a demon taking control of someone's body, the people of ancient civilization simply lumped all together into the grouping of spiritual possession. This means that in many areas where we today think about science and look for a scientific explanation and scientific remedy, the people of the ancient world had no corresponding recourse to science, and they had to turn to help from the supernatural, thus turn to help from religion. This means that we may miss the religious understanding of parts of many stories that include topics such as illness, childbirth, or a thunderstorm. Those religious implications may have been obvious to the original writer and readers, but when we in our cultural framework view these events as simply natural and scientific, we miss the perception of the religious significance that was intended.

    The fact that the creation story is placed at the beginning of Genesis, the first book of the Bible, is significant. But it is the religious significance that should be sought, not the historical or scientific significance. I hold that the story in the book of Exodus is at the heart of the Old Testament, and thus the heart of the Christian Bible, and the book of Genesis is just prologue to Exodus. Exodus is the story of God's revelation to Moses, the great rescue out of Egypt, the establishment of the Sinai Covenant, and the important theology associated with social obedience and with religious symbolism attached to food sacrifice and the tabernacle. Therefore part of the question of interpreting the chapters in Genesis is how do the parts of Genesis connect to Exodus? For example, you cannot have Moses take the people of Israel out of Egypt to the promised land unless you first tell the story of how they became a foreign community living in Egypt, and how there was a promise of some other land. You cannot have God choose Israel and set up this nation as a witness for all the rest of the world unless you have the prior promise that God has a plan promised to Abraham that "through your offspring all nations on earth will be blessed" (Gen. 22:18, also 12:3, 18:18, 20:17-18, 26:4, and 28:14). You cannot have God prepare the symbolism of the sacrifice of the Passover Lamb unless you establish that humans are sinful and need God's saving intervention, as narrated in the stories of the Fall into Sin, Cain and Abel, Noah, and the Tower of Babel. In light of that, how might the stories of Abraham and Isaac establishing altars in Palestine connect with the long prescription of the design of the tabernacle with its altars (Exodus 25-37, 30), and how might the story of God's first creation of Temple Earth connect with that sacred tabernacle?

This is why I find either attempt to harmonize the creation story of Genesis 1—with Newtonian science or with more modern science—unsuitable. The goal of understanding Genesis 1 or any other Biblical passage should be to determine the religious message through the proper interpretation of the sacred text. Until the time when we are sure that we know everything there is to know, we should be careful to keep the games of science and religion separate, and we should play each game according to its own rules. We do bring information from each game into the work of the other, but we must be careful not to break the rules. We use the rules of science to determine science, and we bring the fullness of scientific information to the interpretation of the sacred text. We do not force the interpretation of scripture to fit some preconceived dogma of science, but we do use science to help decide what a word means or does not mean in a certain context. And we use the rules of religion to determine the fullness of supernatural revelation, and we bring that revelation to the interpretation and handling of nature. From religion we learn what God desires us to do in caring for nature, which includes animals, plants, and land, and what God desires us to do in using nature to care for human society. It is from the teachings of religion that we have learned not to fear the god of the volcano, nor to fear offending the goddess of the moon with our rockets, nor to fear the spirits of the dead when we investigate or relocate ancient cemeteries. It is from the teachings of religion that we have learned to place limits on medical science that respect the minds and bodies of all humans, even those who are genetically defective, aged and dying, or condemned prisoners. The fact that supernatural revelation has allowed us to demystify such things does not eliminate the domain of the supernatural.

Conclusion

The world in which we humans find ourselves is a mystery to us. We get to spend our entire lives seeking a better understanding of the world that contains us. We seek to explore it in terms of geography and science, in terms of psychology and social science, in terms of political and economic sciences, in terms of art and music and emotions, in terms of love, and in terms of confrontation with the supernatural. Like an athlete at a track meet, we are enrolled in several events, and we compete in all of those events more or less at the same time as we go through the day of the great track meet of life. But proper and successful competition in each of the separate events requires knowing the rules of that particular event, and knowing the proper technique for the specific event. In the great experience of human life we seek to learn truth in many different areas, and thus in many different ways. As human beings we have the magnificent gift of minds, and we have the obligation to think as well as feel in the pursuit of the understanding of life. We ought to develop and use the gift of reason to its fullest extent. But reason is a tool, a servant in the task of living; reason is not the purpose and master of human life. When you turn to contemplating that which is the purpose of life, and that which will serve as the judge of your human life, and that which fills the role of the master of your human life, you find yourself addressing the big questions of origin, and destiny, and who or what is god?

In our human search for truth, we have learned to divide many of the different aspects of life into focused areas, and we have learned to study the separate questions of those areas individually, to learn the truth that they each contain. For clarity of thought, we divide the different aspects of the search for truth into different games, and we learn and refine the rules for each of the separate games. When participating in each of the separate areas, we play by the rules of that particular game. The aim of this book has been to help you sharpen your understanding of the rules of the game of modern natural science, and the rules of the game of revelatory religion. Only when you play by the rules can you cooperatively play with others, and only when you play by the rules can you be sure that your contribution counts, that it has any real meaning. What good does it do to say you ran a mile in world record time, but you cut off the last corner due to mud and cut the distance a little short? Because you broke the rules, you cannot tell if your effort was really a new human achievement, or just some exercise that contributed to a healthy feeling. If you want to learn what you can about the regular processes that govern the world of nature and what that information reveals about the past history and potential future of the world, including areas such as geology and biology, you have to play by the rules of modern science. If you want to really understand the irregular events that intrude into human life and that seem to interrupt the natural processes, the events that suggest that some other mind is bumping you and butting you and trying to get your attention, you have to play by the rules that govern the inspection and interpretation of the supernatural. You have to tune into God's radio frequency to listen to God, and not insist that he tune in to you. If God has spoken, what did he really say? Only when you properly play by the rules in both the games of science and religion can you be sure that you are perceiving the truth that each has to offer.

www.ingramcontent.com/pod-product-compliance
Lightning Source LLC
Chambersburg PA
CBHW052100300426
44117CB00013B/2217